UNDERSTANDING FAITH

Understanding Jainism

UNDERSTANDING FAITH

SERIES EDITOR: PROFESSOR FRANK WHALING

Also Available

Understanding the Baha'i Faith, *Wendi Momen with Moojan Momen*
Understanding the Brahma Kumaris, *Frank Whaling*
Understanding Buddhism, *Perry Schmidt-Leukel*
Understanding Chinese Religions, *Joachim Gentz*
Understanding Christianity, *Gilleasbuig Macmillan*
Understanding Hinduism, *Frank Whaling*
Understanding Islam, *Cafer Yaran*
Understanding Judaism, *Jeremy Rosen*
Understanding Sikhism, *W. Owen Cole*

See www.dunedinacademicpress.co.uk
for details of all our publications

UNDERSTANDING FAITH

SERIES EDITOR: PROFESSOR FRANK WHALING

Understanding Jainism

Lawrence A. Babb

Willem Schupf Professor of Asian Languages and Civilizations
and Professor of Anthropology, Emeritus,
Amherst College, Massachusetts

DUNEDIN

EDINBURGH ◆ LONDON

First published in 2015 by
Dunedin Academic Press Ltd

Head Office
Hudson House, 8 Albany Street,
Edinburgh, EH1 3QB

London Office
352 Cromwell Tower, Barbican,
London, EC2Y 8NB

ISBNs
Paperback: 978-1-78046-032-1
ePub: 978-1-78046-535-7
Kindle: 978-1-78046-536-4

British Library Cataloguing in Publication Data A catalogue record for this book is available from the British Library

Typeset by Makar Publishing Production, Edinburgh
Printed by CPI Group (UK) Ltd., Croydon, CR0 4YY
Printed on paper from sustainable resources

FSC www.fsc.org
MIX
Paper from responsible sources
FSC® C013604

Contents

List of Illustrations

Romanisation and Pronunciation

Because most of the terms from Indian languages appearing in this book are normally given with diacritical marks in other works on Indian religions, I have given them with diacritical marks here. The exceptions are modern place names (such as Jharkhand instead of Jhārkhaṇḍ) and the names and titles of individuals from the modern period (such as Acharya Shantisagar instead of Ācārya Śāntisāgar). However, the authors of modern works in Hindi are given with diacritical marks (such as K. Kāslīvāl). Most of the specialised terms used in *Understanding Jainism* are from Sanskrit or Prakrit, and I have normally written them with the medial and final short 'a', which is pronounced in these languages but is not pronounced in Hindi or Gujarati. Thus, the word *mokṣa* (meaning liberation) is given in that form, not as *mokṣ*, and the same is true of almost all words that have a context in classical languages. There are a few exceptions, such as Brāhmaṇ rather than Brāhmaṇa and Jain rather than Jaina. Terms that have a mainly modern context (such as caste names) are given in their modern form. Plurals of italicised words from Indian languages are indicated with an added non-italic 's'.

The system of transliteration used here is the same as the one employed by R. S. McGregor (1992) in his *Oxford Hindi-English Dictionary*. Following in the order of the vowels in the Devanagari writing system:

a 'u' as in 'but'

ā 'a' as in 'father'

i 'i' as in 'bit'

ī 'ee' as in 'meet'

u 'u' as in 'pull'

ū 'oo' as in 'fool'

e 'ay' as in 'pay'

o 'o' as in 'vote'

ai 'ai' as in 'aisle' (but in modern Hindi 'ay' as in 'say')

au 'ow' as in 'cow'

Many consonants appear in two forms, underdotted and without underdot. With the underdot (ṭ, ṭh, ḍ, ḍh, ṇ) they are pronounced with the tip of the tongue touching the roof of the mouth. Without the underdot they are pronounced with the tongue touching the back of the upper front teeth. The underdotted ṛ is pronounced as 'ri'. Aspirated consonants (kh, gh, ch, jh, ṭh, ḍh, th, dh, ph, bh) are pronounced with a puff of air following the consonant sound, as in 'redhead'. Although 'ś' and 'ṣ' are separate sounds in Indian languages ('ṣ' is palatalised as are the other underdotted consonants), they may both be pronounced as the 'sh' in 'sheet'. The consonant 'c' is pronounced 'ch', so that *ācārya* is pronounced 'āchārya'. The underdotted 'ṃ' represents a nasalisation shaped by the consonant it precedes. The nasal consonant 'ñ' is pronounced as it is in Spanish; 'ṅ' is pronounced as the 'n' in 'bring'.

Acknowledgements

Two individuals have played indispensable roles in the preparation of this book. John Cort read through the original manuscript and provided invaluable criticisms and suggestions. Surendra Bothara provided similar feedback on portions of the text and always stood ready to help out with the many questions that arose in the course of writing this book. That said, I must admit that I was not always able to do full justice to their suggestions, and they must not be held accountable for any of its flaws. Beyond that, a great many individuals, far too many to mention all, have contributed to my education about Jainism over the years I have been involved with the subject. Thanks are due especially to Dr S. S. Jhaveri (of Ahmedabad), Dr Mukund Lath, Mr Rajendra Shrimal, Mr Milap Chand Jain, Mr Gyanchand Biltiwala, Mahopadhyaya Vinayasagar, Mr Ashok Bhandari, Mr Gyanchand Khinduka and Mr Jyoti Kothari (all of Jaipur). I am also very grateful to Professor Frank Whaling for making it possible for this book to be included in the *Understanding Faith* series.

The Bothra family has been my family away from Amherst for decades, and their home has become my refuge and headquarters in Jaipur. For the loving support they have unfailingly given me they have my deepest gratitude.

My wife Nancy has shared the ups and downs of my life in India and so much more. Beyond that, she has borne the distractions of my labours on this book with patience and fortitude. A thousand thanks.

Introduction

Jainism is an ancient religion of India with roughly 4–5 million adherents in India itself and a small but flourishing overseas community. The 2011 Indian census gives the figure of 4.2 million, but this is likely an undercount because some Jains return themselves as Hindu. This will doubtless strike readers as very tiny relative to the size of India, which indeed it is, but one must not be misled. If the number of Jains is small, their influence in Indian society is very great, far out of proportion to their numerical strength. This is a consequence of the fact that Jains – not all Jains, but most of them in India's North – have specialised in business and business-related occupations. And while it is far from true that all Jains are rich (a national stereotype in India), Jains are among the wealthiest of modern India's religious communities and one of the most influential as well.

In India, and abroad to the extent that they are known at all, Jains are noted for two behavioural traits. One is *ahiṃsā* (non-harm, non-violence), which is an ethic enjoined by their religion as well as a deeply entrenched cultural value. Jain mendicants, in particular, are renowned for the pains they take to avoid harming even the most microscopic of living things. Jains are also well known for the extent to which mendicants, and to an impressive extent laity also, engage in the most demanding ascetic practices, especially fasting. Illustrative of the importance of ascetic practice in Jain life is the fact that some Jains end their lives by means of ritualised self-starvation, and this includes laity as well as mendicants.

Jainism is frequently paired with Buddhism, its far better-known cousin. This makes sense because both traditions came into prominence during the same period of India's history, and they were alike in their mutual rejection of the Vedic traditions that centuries later came to form part of the core of Hinduism. In other respects, however, Buddhism and Jainism are quite unlike, with very different concepts of the ultimate goal of spiritual endeavour and the means to the attainment of that goal.

The term Jain means a follower of a 'Jina', and the term Jina denotes a 'victor' or 'conqueror'. The victory in question, however, is not won on the world's battlefields; instead, it consists of an inner conquest of the desires and aversions that are the root cause of the soul's bondage to the world and its sorrows. And the fruit of victory is not the usual spoils of war but liberation from worldly bondage. More specifically, a Jina is one of Jainism's great mendicant-teachers who, by means of rigorous self-purification, achieves liberating omniscience and – prior to his own final liberation – teaches the truths he has discovered to his followers. These teachers are also known as Tīrthaṅkaras, a term denoting someone who establishes a *tīrtha* (ford, as in a ford across a river). The *tīrtha* in question consists of Jain teachings and the Jain community, which is a community within which these teachings are preserved, transmitted, heard and acted upon. According to these teachings, such communities were, are and will always be established and re-established by an infinite series of Jina/Tīrthaṅkaras, a process that has been going on from the beginningless past and will continue for all of infinite time to come.

Despite its small size, the Jain world is fissured by a number of important sectarian divisions. Of these, one should be mentioned at the outset, because it will be with us throughout the remainder of this book. This is the split between Jainism's two main branches: the Śvetāmbaras and Digambaras. From the standpoint of soteriological doctrine, the difference between them is not great, but they are separated by a wide chasm indeed on a point of monastic discipline, namely whether or not mendicants should wear clothing. The term Śvetāmbara means 'white clad', and refers to the Jain branch in which monks and nuns wear white clothing. Digambara means 'space clad', and the monks of this branch (not the nuns) wear nothing. In general, the Śvetāmbaras predominate in India's North, although there are substantial Digambara communities in the North as well. The Jains of South India are almost all Digambaras. The exact proportion of Śvetāmbaras versus Digambaras is not known, but Digambaras are probably somewhat less than half the total.

Understanding Jainism is an introduction to Jain belief, practice and tradition. Our starting point is the history of Jainism and its place in the history of Indian religions. We then turn to Jainism's most fundamental teachings about the nature of reality and the human situation. An understanding of these doctrinal basics provides an entry into an exploration of how Jainism's formal teachings are and are not embodied in mendicant and lay ways of life. We then set Jain belief and practice in a wider context of Jain cosmography, geography and biology, and we turn finally to the ways in which Jainism is linked to the identities of social groups in India. The groups in question are not those idealised in Jain sacred writings, but ones highly relevant to the actual lives of Jains.

We begin with a critical episode in Jain history as seen from the standpoint of Jain tradition.

Chapter 1

Charter

The heart of this chapter is a story that one must know in order to make the slightest sense out of Jain beliefs and tradition. It is about the conception, birth, world renunciation, enlightenment and final liberation of the most recent of the Tīrthaṅkaras in our part of the cosmos. His given name was Vardhamāna, but he is best known to history as Mahāvīra, a title and not an actual name, meaning '*mahā* (great) *vīra* (hero)'. According to Jain teachings, he was the last of the twenty-four Tīrthaṅkaras of our cosmic epoch (one of moral and physical decline) and our small section of the terrestrial world.

It is a narrative that can be read in two different ways. On the one hand, the history of Jainism, as modern historians understand this term, largely begins with his life. We know that Jainism had a prehistory, and we understand also that Mahāvīra had a historically verifiable predecessor named Pārśva. However, Jain tradition enters clear historical visibility only after his arrival on the scene. He is often represented as the founder of Jainism. This is a claim that is neither fully true nor fully false, and we shall return to it anon. For the present it is enough to say that to place Jainism in its historical context one must begin with Mahāvīra's life. On the other hand, the narrative also possesses a very different kind of importance within Jain tradition. To the Jains, its importance is normative as well as historical. It is an example of a 'charter' narrative, by which is meant that it provides a pattern and source of legitimacy for key elements of Jain belief and practice.

I have abstracted the version presented here from a text called the *Kalpasūtra*, believed to have been composed by Bhadrabāhu, an ancient sage who is said to have died 170 years after Mahāvīra's death. Among the materials it contains are accounts of the lives of four Tīrthaṅkaras, and of these the biography of Mahāvīra is much the longest and most detailed, which is in keeping with his importance in Jain tradition. I have chosen this text (drawn here from Vinayasāgar, 1984) for two reasons. First, it is one of the two earliest sources of his life we have, probably dating from the second or first century BCE. The other is the latter part of the *Ācārāṅgasūtra* (Jacobi, 1884), which is somewhat earlier but belongs to roughly the same period. And second, this text, though ancient, plays a very important role in the present-day ritual culture of image-worshipping Śvetāmbara Jains. Its recitation is one of the main events in the celebration of Paryuṣaṇa, the year's most important religious observance, and its role in the ceremony testifies to its importance as a repository of the all-important story of Lord Mahāvīra's life.

It should be noted that the Digambaras tell the story of Mahāvīra's life somewhat differently, and I indicate some of the major differences in both my retelling of the story and the historical chapter that follows.

The Lord's life

At midnight on the sixth day of the bright fortnight of the lunisolar month of Āṣāḍh [June/July], *the soul of Mahāvīra descended to this earth from the heavenly region where he had lived for many eons as a god. He entered the womb of a Brāhmaṇ woman (i.e., a woman belonging to the priestly class) named Devānandā who lived in the Brāhmaṇ sector of a town called Kuṇḍagrāma. Then, as she was lying on her bed, half asleep and half awake, she experienced fourteen wonderful dreamlike visions. She saw an elephant, a bull, a lion, the goddess Lakṣmī* [the goddess of prosperity] *being anointed, a garland, the moon, the sun, a flag on a golden staff, an urn, a lotus pond, the milky ocean, a vast celestial vehicle, a heap of jewels and a smokeless fire. When she awakened her husband and told him about the visions, he replied that they were highly auspicious and foretold the birth of a son who would possess every virtue and perfection and who would master every branch of learning* [a career appropriate for a Brāhmaṇ].

Now, Indra, the mighty king of the gods, who was seated in his council and who keeps constant watch on the terrestrial world below, came to know that Mahāvīra had entered the womb of a woman living in the continent of Bharata (our part of the terrestrial world). Rejoicing, he rose from his throne and paid obeisance to Tīrthaṅkara-to-be. But after returning to his throne, he realised that something was amiss, for it was quite unsuitable and indeed impossible for a Tīrthaṅkara ever to be born of a Brāhmaṇ woman, because such an august personage must be born in an aristocratic Kṣatriya lineage (i.e., belonging to the class of warriors and rulers). He thereupon arranged for Mahāvīra's foetus to be removed from Devānandā's womb and conveyed to the womb of Triśalā, a noble woman of Kṣatriya lineage, also pregnant at the time, and for the foetus in Triśalā's womb to be placed in the womb of Devānandā [this episode is not accepted by the Digambaras].

On that very night, which preceded the dawn of the eighty-third day after Mahāvīra had entered Devānandā's womb, Triśalā, while lying half asleep, saw the same series of auspicious dreamlike visions that Devānanda had seen before, but even more magnificent. As Devānandā had done, she awakened her husband, who was a king named Siddhārtha, and told him of the visions. Siddhārtha responded that the dreams foretold the birth of a son in nine months and seven and one-half days. He would bear every virtue and perfection and would become a mighty warrior and king [a career appropriate for a Kṣatriya].

Two days later, having ordered his audience hall to be prepared for the occasion, Siddhārtha recounted Triśalā's visions to the interpreters of dreams. The visions foretold, they said, the birth of a son who, in his manhood, would become a great king and warrior and would rule the world's four-quarters as a cakravartin [universal emperor]. *But then they added that his son might alternatively become a* 'cakravartin *of* dharma', *a Tīrthaṅkara, a victorious spiritual warrior who would lead the world as a great teacher.*

As Triśalā's pregnancy ripened, the gods filled Siddhārtha's coffers with treasures and Siddhārtha's clan, the Jñātṛs, began to flourish, as did the whole kingdom. Because of this, the expecting parents decided that they would name their son Vardhamāna, the Increasing One.

While in the womb, Mahāvīra made no movements in order not to cause his mother pain, but then Triśalā, fearing that something was wrong, began to fret. Even though he was still in the womb, Mahāvīra was fully aware of his mother's feelings, and to assuage her anxiety he made a slight movement that she felt to her relief and joy. At that moment, Mahāvīra vowed not to renounce the world while his parents still lived out of respect for their feelings. [That he so vowed is not accepted by the Digambaras.]

Lord Mahāvīra was born at midnight on the thirteenth day of the bright half of the lunisolar month of Caitra [March/April]. *At the moment he was born, huge celebrations erupted among the gods in their heavens and in the kingdom below. The gods showered the palace with wealth and anointed the newborn boy. Siddhārtha ordered the kingdom's prisoners to be released, and the entire town celebrated the birth of Siddhartha's son and heir for ten whole days.*

Lord Mahāvīra spent thirty years of his life as a prince, after which he renounced the world. [The Digambaras say he decided to renounce in his thirtieth year with the reluctant consent of his mother. The Śvetāmbaras believe that he waited until after his parents had died.] *He gave up everything: his wealth, his kingdom, his armies, everything he possessed, great or small. Then, at the beginning of winter on the tenth day of the dark fortnight of the lunisolar month of Mārgaśīrṣ* [November/December], *having shed all his possessions, he left his home on a litter followed by a great congregation of gods, men and even demons. When he arrived at an Aśoka tree in a nearby park he dismounted, discarded his ornaments, pulled his hair out in five handfuls, and vowed henceforth to take only one waterless meal out of six. Wearing only a single cloth, he became a homeless mendicant, and the cloth he gave up after a year and a month on the road.* [The Digambaras say he was nude from the start.]

For twelve years Mahāvīra wandered as a homeless mendicant and bore the many discomforts of such a condition with complete equanimity. Except for the annual four-month rainy season retreats when he would stay in one place, he never spent more than a night in any village, and never more than five nights in a town. He subjected himself to the most extreme of privations. In this way he freed himself of all destructive feelings and achieved a condition of tranquillity and detachment from the world and all things of the world.

In the thirteenth year of his homelessness, on the tenth day of bright fortnight of the lunisolar month of Vaiśākh [April/May], *Mahāvīra, who had been eating only one waterless meal every three*

> *days, and while meditating under a tree, and while exposing himself to the relentless rays of the summer sun, attained omniscience and all-seeingness, thus becoming a Jina (i.e., a Tīrthaṅkara), a spiritual conqueror. As an omniscient being, he knew everything about all beings – gods, humans, and the denizens of hell – their thoughts, their feelings, their conditions of life, their pasts and futures. Nothing was concealed from him.*
>
> *After his attainment of omniscience, Lord Mahāvīra lived on for nearly thirty years more, wandering from place to place except during the rainy season retreats. He died – which is to say, he attained complete liberation from the bondage of birth, death, and decay – in his seventy-second year at the dawn of the fifteenth (moonless) night of the dark half of the lunisolar month of Kārttik* [October/November]. *On the night of his liberation, the gods made a glorious display of their joyful celebrations.*

The text then goes on to enumerate the followers that Mahāvīra left behind: 14,000 monks (led by Indrabhūti Gautama, his chief disciple), 36,000 nuns, 59,000 laymen, 318,000 laywomen, 300 disciples who had total command of the scriptures, and many other followers in several categories at various spiritual capacities and levels of attainment. The remainder of the *Kalpasūtra* contains much shorter biographical sketches of three other Tīrthaṅkaras: Pārśva, Ariṣṭanemi and Ṛṣabha. It also includes a chronology of the twenty-four Tīrthaṅkaras of our declining epoch and area of the terrestrial world, which seems to be the earliest reference to the full list. The text then provides an account of the mendicant lineage established by Mahāvīra, beginning with his eleven chief disciples. The later Śvetāmbara monastic community regards itself as descended by disciplic succession from one of these, Sudharma, and the text describes the development and ramification of the lineage of his spiritual descendants. (The Digambara disciplic pedigree is different; see Wiley, 2004, p. 151.) It concludes with instructions for the conduct of mendicants during the four-month rainy season retreat and the holy days of Paryuṣaṇa together with a brief discussion of types of microscopic living things.

Under many layers of embellishments (of which my brief summary gives little true idea), we should note that this narrative can be further reduced to certain basics. First, the text focuses on five essential events in Mahāvīra's biography: his conception, birth, renunciation of the world, attainment of omniscience and final liberation. These events are known as the five *kalyāṇakas* (auspicious events), and Jain teachings maintain that, although specific details of the Tīrthaṅkaras' biographies differ, these identical auspicious events occur in the lives of all of them. The five *kalyāṇakas* are definitive of what a Tīrthaṅkara actually is, and they are a profoundly important element in Jain belief and ritual culture.

Second, and equally important, the narrative connects the far distant past of Mahāvīra's life to succeeding periods by telling how he created a fourfold social order (known as the *caturvidha saṅgha*) consisting of monks and nuns, laymen and laywomen, and established a line of disciplic succession by means of which his teachings could be preserved and propagated. That social order is seen as the precursor of the Jain community that exists today, and the line of disciplic succession invests the existing Śvetāmbara mendicant community with a spiritual authority ultimately devolving from Mahāvīra himself.

Chapter 2

In History

We now shift from the perspective of charter narrative to that of history as understood by modern historians. Let us start by noting that there can be no doubt about the historicity of Mahāvīra. The unanimity within the Jain fold that he once lived and taught in ancient India and the fact that he is known to have been a contemporary of the Buddha leave little room for doubt about his actual existence. According to Śvetāmbara tradition, he was born in 599 BCE and died in 527, whereas the Digambaras give the dates as 582 to 510 (Wiley, 2004, p. 6). However, recent scholarship has established the date of the Buddha's death as between 411 and 400 BCE, and Mahāvīra and the Buddha were contemporaries, so Mahāvīra's dates should probably be pushed forward by a century or so (Dundas, 2002, pp. 24–5).

The supposed place of his birth, Kuṇḍagrāma, is said to be a village in Bihar named Basukund today. In Mahāvīra's day it was a satellite community of the city of Vaiśāli, capital of the Vaiśāli state and one of the most important cities of northern India. The *Kalpasūtra* portrays his father, Siddhārtha, as a king, but all we can be reasonably sure about is that he was a member of the Kṣatriya class of rulers and warriors. How he spent his first thirty years is in dispute. The Digambaras say that he was averse to worldly pleasures from the start and was chaste throughout his life. The Śvetāmbaras say that he married a woman named Yaśodā and had a daughter. But both sects agree on the essential thing: that he renounced the world and all of the attachments of his former existence and became a homeless wanderer.

We must now ask what such an act meant in context, and to answer this question requires an excursion into Indian history and the history of Indian religions. We begin with some essential early Indian history.

Indo-Aryans

At the time of Mahāvīra's birth and career, North Indian civilisation had entered a phase of rapid urbanisation and intellectual and spiritual creativity, but this was not the first urbanisation of India. Incredibly, nearly two thousand years previously there had developed an earlier city-based civilisation that flourished in the valley of the Indus river and adjacent regions. Almost certainly, the peoples who built this civilisation spoke a Dravidian language, which is the language family of peninsular South India today. The archaeological evidence points to *c.*2600–1900 BCE as the period of its urban maturity, and for reasons that are uncertain it underwent a period of decline and disintegration after 1900 BCE. The peoples who created the Indus Valley civilisation did not simply vanish; rather, they and their culture unquestionably left an imprint on the development of later Indian civilisation, although the depth of that imprint is a contested matter.

By *c.*1500–1400 BCE, but perhaps as early as 1700, a new and culturally very different wave of peoples had begun to appear in the Upper Indus Valley in the area known as Punjab today. These Indo-Aryans (as they are usually called) were a pastoral and semi-nomadic people who spoke archaic dialects of Sanskrit, a language belonging to the Indo-European language family. Their arrival must have taken the form of a slow sifting of small groups through the mountain passes of the north-west frontier. They settled first in Punjab, and it is certain that they mingled and intermarried with peoples they found there. With time, they shifted to an agricultural subsistence base to which their former pastoralism was no longer central, and by *c.*1000 BCE they were pushing southward and eastward into the upper Ganges valley and beyond.

The Vedas

Although there is some archaeological evidence, much of what we know about early Indo-Aryan culture comes from orally transmitted religious texts (only much later reduced to writing) called Veda (meaning knowledge). The Vedas consist of four bodies of materials composed sequentially (though with overlap and interpenetration) from *c.*1500 to 500 BCE, and this period is often called the 'Vedic' period. These materials are available for our perusal today because of a cultural emphasis on their memorisation and accurate recitation as well as mnemonic aids embedded in Vedic pedagogy.

The earliest of the Vedic materials is a collection of 1028 verses called the *Ṛg Veda*. It is the earliest of four *saṃhitās* (collections) of material that together comprise the earliest of the four bodies of Vedic materials just mentioned. These verses, composed in a type of Sanskrit called Vedic Sanskrit, are the earliest materials we have in Sanskrit. While their composition must have occurred over hundreds of years, the earliest of them probably dates from *c.*1500 to 1400 BCE.

The verses of the *Ṛg Veda* were intended to be recited or sung during ritual sacrifices. These sacrifices, which often involved the killing of animals, were performed on behalf of the deities of the Indo-Aryan pantheon, many of which embodied the natural forces that were so crucial to the lives and livelihoods of these peoples. There were no temples in those days and no carved images of deities; instead, the sacrifices were conducted on temporary altars. The basic idea was that the sacrificial fire would carry the offering upward to the gods above. The sacrifices were presented in the hope that the offerings and the verses of praise would persuade the deities to aid the supplicants in their worldly endeavours. The ideas of rebirth and liberation of the soul, so important in later Indic thought, were entirely unknown to the people of the early Vedic period.

The remaining three portions of the Vedic corpus (leaving aside the other three *saṃhitās*) span the period *c.*1000–500 BCE and after. These texts reflect an era in which the Indo-Aryans and their civilisation had moved out of the Punjab and were pushing down the Ganges valley. As they travelled and settled, they developed ever more complex forms of social life, culminating in the rise of states and cities and an increasingly complex religious culture. The fourth and latest of these bodies of material are the Upaniṣads, justly famed for the speculative philosophy they contain. The earliest of them probably dates from *c.*600 BCE. The Upaniṣads are the earliest texts in which we encounter the theory of transmigration of the soul, to which we return shortly.

Vedic social theory

The Indo-Aryans conceptualised their society as a stratified social order consisting of four *varṇa*s (hereditary classes), each with its own function seen as complementary to the functions of the others. This conceptualisation, which probably matured at some point after 1000 BCE, unquestionably reflected social reality to some extent, but it was an idealisation of society, not an ethnographic description. It was expressed in a ritual idiom utilising the concept of the sacrifice, so central to the Indo-Aryans' religious culture. A famous hymn from the tenth (and late) book of the *Ṛg Veda* (10.90) describes the creation of the world as a sacrificial act. The gods, the hymn tells us, made a sacrificial offering of Puruṣa, a sort of universe-sized, manlike entity. Along with everything else in the world, the social order emanated from this primordial rite. From the Puruṣa's mouth came the Brāhmaṇs, whose social task was to preserve and transmit the Vedas and to serve as priests in the sacrifice. From his arms came the Rājanyas (later called Kṣatriyas), the class of rulers and warriors. Ordinary tribesmen, at first called Viś (later Vaiśyas), came from his thighs. Initially, this was the class of ordinary farmers, but in time it came to be identified mainly with traders. Finally, his feet became the Śūdras, whose function was to 'serve' the upper three classes. The upper three classes came to be known as 'twice-born' because males of these classes were initiated in a ritual seen as a second birth, and supposedly only the twice-born were allowed to hear the Vedas. External to this ideal social system were groups deemed too defiled to be included in the social order; they were the ancient equivalents of the 'Untouchables' of later times.

Crucial to this concept of society was the dyad at the top, the Brāhmaṇs and Kṣatriyas, and their relationship was also conceptualised in a sacrificial idiom. For the welfare of the whole social body, the performance of sacrifices was held to be essential, but only the Brāhmaṇ priest could actually perform these rites. In turn, sacrifices (or at least the great ones) were sponsored by Kṣatriya rulers, whose legitimacy as rulers was secured by sacrificial sponsorship. Within the compass of this image, the Brāhmaṇ, with direct access to divine power, can be seen as the apex of the system, with the Kṣatriya always second in rank (Dumont, 1970), but in fact the issue of which of these classes is superior has always been contested territory (Trautmann, 1981, pp. 283–8).

This entire scheme reflected a rural social order in which tribe and lineage and a simple trichotomy of priests, warrior-rulers and farmers (with servants added) were the foremost social formations. In it, religious life was simple and mostly focused on the here and now of crops, herds and the fortunes of war.

Mid-millennium and the changing religious landscape

Beginning in roughly the sixth century BCE, the older rural and village-based tribal order, centred in Punjab, gave way to profound economic, social and political changes. The venue of change was the Gangetic plain where the centre of gravity of Indo-Aryan civilisation had now shifted. This was a period marked by the growth of cities – a reappearance of urban life in India after a hiatus of more than a thousand years – and the formation of states, which ultimately gave rise to an empire (the Nanda empire) in the fourth century BCE. These developments were fuelled by caloric input from agricultural surpluses made possible by the fertility and abundant rainfall of the middle and lower Ganges. A key development was the advent of irrigated rice cultivation, which was probably under way by *c.*500 BCE.

Economic growth encouraged the emergence of craft and other non-subsistence-oriented occupational specialities, and also made possible the emergence of bureaucracies and other social instruments for the organisation and application of state power. In addition, the umbrella of protection extended by the state gave rise to increased trade, which in turn led to the rapid growth of a large trading and banking class that became a powerful force in social and political life. It has been argued that this was a period in which something like 'individualism' becomes historically visible for the first time in India (Olivelle, 1993, p. 58). As is generally true of simpler forms of social life, personal identity in early Vedic society was largely based on group membership: local community, lineage, family. Urban life, however, corrodes these corporate sources of personal identity, and in the flux of urban life persons can exercise choice in determining what they will do and become. This is true in many departments of life, but especially in trade, where so much depends on one's personal strength of will, initiative and industriousness.

This new social world provided a context in which religious life underwent a radical refocusing, at least for some. The older Vedic outlook reflected the mentality of a rural people who were not much given to reflection about life's deeper mysteries, and whose worldview was founded on a rather optimistic faith in the simple efficacy of their sacrificial transactions with their gods. Very different was the religious outlook that emerged in the mid-first millennium. We may reasonably assume that the religious life of most people continued in old patterns, but among the thoughtful few (how many we have no way of knowing) there emerged an unprecedented pessimism about the human situation and a sharpened focus on the destiny of the individual.

It is hard to know what exactly gave rise to this trend, but the social and cultural uncertainties brought about by rapid social change must have played a role. The slow rhythms of village life and the certainties encouraged by the predictability of such a way of life were now replaced by a less predictable social world in which disparate groups needed to find new ways engaging with economic life and of interacting with each other in crowded urban settings. This was a context in which old assumptions about what the world was like were challenged as they had never been before.

Among the religious virtuosi of this period there arose the idea that creaturely existence itself is a kind of all-encompassing misfortune. A key ingredient in this view of things was the idea of the transmigration of the soul, or self, of the individual, a concept that has remained at the heart of Indic religions from those days until now. It came to be believed that death is merely the entryway into rebirth in a new body, which is inevitably followed by death and rebirth again in an endless cycle (called *saṃsāra*). One might, of course, take the bright (if naïve) view that the cycle of *saṃsāra* was one of perpetual beginnings, but that was not the way it looked to the religious thinkers of mid-first millennium BCE India. To them it was an ongoing calamity, not perpetual beginnings but ceaseless separation and death (*punarmṛtyu*, 'redeath'). Even a fortunate rebirth was a misfortune when seen in this light, for the most intense pleasures of life were transient, and the best of lives would end in death. To these thinkers, the round of rebirth was evil in its essence, a form of bondage, and their spiritual energies were focused on finding ways of escaping it, of achieving liberation from what was often called 'the jaws of death'.

Coupled with the idea of *saṃsāra* was the emergence of another concept, that of *karma*, that has also remained at the core of Indic religious thought ever since. The term *karma* in its simplest and basic denotation means simply 'action', but with the implication that action has consequences, that it bears *phala* (fruit), good or bad. The original context for this idea was the old Vedic sacrifice; *karma* was the ritual action of the sacrifice, which was intended to bear fruit for the sacrificer. In time, this idea was abstracted from its initial context and fused with the idea of transmigration, which yielded the notion that action in general led to consequences that must be experienced by the actor, if not in this life then in some subsequent birth. The nature of the consequences depended on the nature of the act: good acts yielded good results, while bad produced bad ones.

As commentators have often pointed out, this intellectual construct represents a nearly airtight solution to the problem of making moral sense of good and ill fortune. If a virtuous person meets with misfortune, or an evil person with apparently undeserved blessings, this can be explained on the basis of behaviour in some previous birth, with the corollary that the balance will be righted in an unknowable future. Obviously, the presumption of amnesia about previous births shields the system from any possible disconfirmation. But it must be also borne in mind that, while an individual might experience good as well as bad in the course of a world-career, the cycle as a whole is ultimately one of sorrow and death; thus liberation from its clutches – not the mere seeking of rewards for good behaviour – must necessarily be the highest religious goal.

These ideas gave rise to a range of religious innovations that diverged radically from the older Vedic sacrificial tradition. The basic goal was to escape *saṃsāra*'s grip, but the means of escape could not be the performance of rituals, nor indeed action in any normal sense, because action is *karma*, and *karma* and its effects perpetuate rebirth. That is, action, good or bad, could hardly be the solution to the problem because it *is* the problem. Instead, liberation from the cycle of birth and death would require interfering in some fashion with the karmic process itself.

Responses to our worldly situation, thus conceived, were various, but there was at least one common thread. This was the idea that seeking liberation required turning one's back on worldly life and disengaging from the web of action and its consequences that is *saṃsāra*'s snare. Whether this idea originally came from within the Brāhmaṇical tradition or from outsiders to that tradition is a contested matter; what is clear is that the conviction that the proper response to the world is ascetic withdrawal took place both outside and inside the Brāhmaṇ fold. Patrick Olivelle (1993, especially pp. 58–63) has made the persuasive suggestion that, among Brāhmaṇs, it evolved among urban Brāhmaṇs whom city life had shaken loose from older orthodoxy, presumably while their country cousins continued in their time-honoured ritualistic ways. Outside the Brāhmaṇ fold, world renunciation became central to various ascetic traditions, apparently originating mainly in the Kṣatriya class. Non-Brāhmaṇ mendicants came to be distinguished from Brāhmaṇs as Śramaṇas (meaning strivers or exerters), so named because of the strenuous nature of their ascetic spiritual quest. Mahāvīra was one of them.

It must be stressed that, whether among Brāhmaṇs or Śramaṇas, world rejection was associated with an extremely varied range of soteriological systems about which it is very difficult to generalise. A well-known Brāhmaṇical view, first (and famously) emerging in the Upaniṣads and familiar to many Western readers in a later formulation called *advaita vedānta*, stressed the role of a special kind of 'knowing' (as opposed to 'doing') as the path to liberation. According to this view, behind the apparent diversity and ceaseless change of the normally experienced world was a deeper unity that was beyond change and thus beyond death. This ultimate reality was called *brahman*, and one's true *ātman* (soul or self) was in fact one with *brahman*. Liberation was to be gained by turning one's mental gaze inward and grasping this truth directly, without the mediation of senses and the normal categories of perception, language and thought. Such a knower was liberated in life (i.e., from a false understanding of his or her relationship with the cosmos), and after death would not be reborn again.

Among the non-Brāhmaṇical schools, two were of special importance: Buddhism and Jainism. Both the Buddha and Mahāvīra appear to have accepted the basic doctrines of *karma* and transmigration as uncontested premises, not as new or novel proposals. Each promoted the life of mendicancy, although the Buddha stressed this far less than Mahāvīra and the Jains. Beyond this, there was very little similarity between these two systems. Although it is conventional to place the two traditions in a single category as 'heterodox' as opposed to the 'orthodox' Upaniṣadic systems, this can be very misleading. Buddhist and Jain tradition incorporated very different understandings of the nature of the self and liberation; indeed, they were as different from each other as either was from the Upaniṣadic system just described.

The turn away from life in the world and its rewards represented a repudiation of values central to the old order. At the core of the Vedic socio-religious system was the figure of the householder, the ideal man. Far from rejecting the world, he embraced it; he was a married man who created wealth, progeny (the source of his immortality) and – above all – offered sacrifices to the gods. This last point is critical, because performance of the sacrifice, a householder's highest duty, required the presence of a wife, and thus – virtually by definition – the path of world renunciation was closed to him. But from the perspective of the ascetic movements, the householder's path was second-rate and worse. From the standpoint of the goal of liberation as the renouncers came to understand it, there was no way that sacrificial rites could possibly release one from the jaws of death, for sacrificial rites were the very paradigm of *karma*, which was the source of worldly bondage.

In the ascetic traditions, Brāhmaṇical or non-Brāhmaṇical, the ideal of the householder was thus turned on its head. The ego-ideal was now precisely the man who rejected the claims of family, who did not marry and who did not perform sacrificial rites. Or at least he did not perform such rites externally, for ascetic practice was seen as a way of internalising the rite, with austerity and self-inflicted discomfort kindling a fiery power within his own body. The ascetic movements thus pushed the householder off his former pedestal (Olivelle, 1993, pp. 64–7). Homeless wandering, poverty and celibacy became the crucial conditions for escape from death. Celibacy had a special importance, for marriage was the gateway to family life, and family life was the gateway to all of the social and economic ties that, from the ascetic's standpoint, were fatal to the quest for immortality.

This was the intellectual context in which Mahāvīra came of age spiritually and ultimately renounced the world. In turning his back on the world, he was certainly not innovating, for the roads and back lanes of northern India in those days surely saw the comings and goings of many similar figures. His distinctiveness, rather, lay in the doctrines he taught, the extent to which he stressed the importance of austerity and, of course, the success of his teachings in attracting a following.

Mahāvīra's mission

It should be understood that, from the perspective of Jain tradition, Mahāvīra was not the founder of Jainism. The Jains maintain that the teachings of Jainism are eternal and that they are periodically rediscovered by certain extraordinary individuals – the Tīrthaṅkaras or Jinas – who teach these timeless truths and establish the communities that preserve and put them into practice. Because the cosmos is uncreated and will never end, these teacher-establishers are infinite in number and come and go for all of infinite time.

Further, we have noted already that, even if we adopt a more strictly historical point of view, Mahāvīra had at least one historically identifiable predecessor. According to Jain teachings, the twenty-third Tīrthaṅkara of our declining epoch and corner of the world was Pārśva, and he, too, was almost certainly an actual historical person. According to the *Kalpasūtra* he lived 250 years before Mahāvīra. He is said to have been born in Varanasi, to have renounced the world and achieved omniscience, to have founded a mendicant lineage and to have died after a one-month total fast (i.e., no food and no water) at the age of a hundred at Sammet Shikhar in the current Indian state of Jharkhand.

The nature of Pārśva's influence on Mahāvīra is an obscure matter. It seems nearly certain that Nigaṇṭhas (unattached), as the Jains were then called, already existed when Mahāvīra was born. According to the *Ācārāṅgasūtra*, the oldest source we have on his life, Mahāvīra's parents were actually followers of Parśva (Book II, 15.16; Jacobi, 1884, p. 194). Furthermore, Buddhist sources indicate that prior to the Buddha's lifetime there lived mendicants who followed a tradition of four monastic vows. These were presumably the vows that Jains (in the Śvetāmbara interpretation) associated with Pārśva: no violence, no falsehood, no theft, no possessions (but see Bronkhorst, 2000, pp. 515–17 on translation problems pertaining to the fourth vow). If nothing else, this seems to show that there was indeed a Jain ascetic community in existence before the Buddha's lifetime (and thus before Mahāvīra). Mahāvīra, then, is said to have added a fifth vow to the list, namely the vow of chastity. Putting these contentions together one can surmise that Mahāvīra must have had some kind of direct relationship with Pārśva's ascetic lineage, and was perhaps initiated into it, in which case he would best be regarded as a 'reformer' of a pre-existing tradition (K. C. Jain, 2010, vol. 1, p. 5).

The *Kalpasūtra* tells us that Mahāvīra wandered for twelve years after his renunciation of the world, facing innumerable discomforts and difficulties with equanimity, and we can be sure that he met many similar figures who had also taken up the life of homeless wandering. As far as we know, however, he developed a prolonged relationship with only one of them. This was an ascetic named Gośāla, who was the leader of a mendicant sect, now extinct for many centuries, called Ājīvika (on which see Basham, 1951; Dundas, 2002, pp. 28–30). Members of this group were nude and famed for their deterministic and fatalistic views. It seems possible that Mahāvīra was inspired by Gośāla to relinquish his own clothing, because followers of Pārśva probably wore clothing. The Śvetāmbaras maintain that Gośāla became a disciple of the not-yet-omniscient Mahāvīra, accompanied him for a total of six years, and in the end magically attacked and split from him, falsely proclaiming himself to be fully enlightened. The Digambaras, who believe that Mahāvīra maintained a vow of silence during his twelve-year pre-enlightenment wanderings, have no tradition of such a relationship.

As we know, Mahāvīra and the Buddha were contemporaries, but apparently they never met. We do know that during his pre-enlightenment wanderings the Buddha-to-be tried the path of severe austerities – precisely the sort of path for which the Jains are famous – and found it wanting. In the end, he concluded that such severe self-mortification led only to severe emaciation, not to enlightenment. He turned away from such an approach and advocated a 'middle way', i.e., neither excessively permissive nor excessively ascetic.

For his part, Mahāvīra never deviated from the strict ascetic path. According to Jain teachings, he attained omniscience during the thirteenth year of his homeless wandering. From the Jain standpoint, this is the crux of the matter. Each of the five auspicious events in the life of a Tīrthaṅkara has its own importance, but the attainment of omniscience is both the culmination of the previous three and the event that sets the stage for everything to come. It is at this point in his life that he becomes an omniscient teacher and establisher of a community of mendicant and lay followers. His own liberation is but a matter of time, to be achieved with the elimination of certain slight karmic residues, but in the meantime he pursues his teaching mission.

The *Kalpasūtra* has relatively little to say about Mahāvīra's nearly thirty years of teaching. It does, however, provide a list of places where he spent his rainy season retreats. To the extent that it is accurate, it suggests that the area of his wandering and teaching was mostly within the boundaries of the present-day states of Bihar and Jharkhand.

Jain tradition tells us that Mahāvīra taught (as do all Tīrthaṅkaras – past, present and future) from the apex of a special structure the gods created for his first sermon and recreated for subsequent sermons as well. Such a structure is called *samavasaraṇa* (place of universal resort), so named because his sermons were attended by deities, humans and even animals. It is a circular structure with the Tīrthaṅkara and his disciples at the top and the highly varied audience distributed among three descending concentric circular galleries. He is seated on a cushion facing eastward and surmounted by a white umbrella – a symbol of royalty. The gods create three replicas of the Tīrthaṅkara so that his sermon can be projected in all four directions, putting an accent on the universality of his teachings.

There are significant differences, however, in the manner in which the Śvetāmbaras and Digambaras represent the Tīrthaṅkara's teaching persona. In general, the Digambaras tend to downplay his human characteristics to a greater extent than the Śvetāmbaras. The Śvetāmbaras believe the post-enlightenment Mahāvīra to have still been a human being, although an extraordinary one; he ate physical food, he excreted and he spoke and preached in a human language, a Prakrit called Ardhamāgadhī. The Digambaras maintain that post-enlightenment Mahāvīra was, in effect, human in form alone; his nourishment consisted of a special divine food that he absorbed and did not eat, and so tenuous was his engagement with the physical world that he sat motionless while delivering sermons that were expressed by means of a divine sound emanating from his body. The sound required translation into human language by his chief disciples before it could be understood by others.

The *Kalpasūtra* tells us that Mahāvīra died and achieved liberation at the age of seventy-two at a place called Pāpā (now said to be the town of Pavapuri in Nalanda District, Bihar) in the scribes' hall of a king named Hastipāla. According to Śvetāmbara tradition, the date of his liberation was 527 BCE; the Digambaras maintain that it was in 510.

His following

We may assume that Mahāvīra's teachings were mediated by a small group of especially devoted followers, for this is typical of the way new religions or dispensations within an older tradition establish a social foothold (see Stark and Bainbridge, 1985, on the sociology of such movements). Jain traditions relating to the general contours of this group point to just such a process, and Śvetāmbaras and Digambaras are in general agreement about overall narrative. That said, we have to remain less confident about the details – the names, the backgrounds and the numbers of followers – for much of this material must have been supplied to the basic narrative in subsequent centuries.

Let us begin with the earliest conversions. According to Śvetāmbara tradition, Mahāvīra made no converts at his first preaching assembly; it was attended only by the gods, who are incapable of renouncing the world, so he did not even preach (for variations between and within versions, see Dundas, 2002, pp. 39–41; P.S. Jaini, 1979, pp. 37–9). He was more successful at his second preaching assembly. It occurred when he came upon a bloody sacrifice. His sermon, delivered there on the spot and on the subject of non-harm, converted three of the attending Brāhmaṇs. They were three brothers belonging to the Gautama *gotra* (roughly, a clan), and the most senior was named Indrabhūti. These three, joined later by eight other Brāhmaṇs, became Mahāvīra's *gaṇadhara*s (chief disciples) with Indrabhūti as the head *gaṇadhara*. In a Digambara version, Indrabhūti, having failed to understand

Mahāvīra's teachings (as imparted to him by the god Indra disguised as a Brāhmaṇ), came to the *samavasaraṇa* sixty-six days after Mahāvīra's attainment of omniscience; there he experienced an instant conversion as a result of the sight of the pillar in front of the assembly. He, in turn, brought his two brothers into the fold, and they were followed by eight other Brāhmaṇs.

We must note before passing on that both Digambaras and Śvetāmbaras believe all eleven *gaṇadhara*s were Brāhmaṇs. The Śvetāmbaras put additional emphasis on their Brāhmaṇhood by placing the initial encounter at an animal sacrifice, which Jains both detest and see as the defining institution of the old Brāhmaṇical socio-religious order. In stating that they were Brāhmaṇs, the narrative makes the point that Brāhmaṇs – who were once the sole possessors of religious authority – were now acceding to the superior wisdom and spirituality of a great non-Brāhmaṇ teacher who belonged to the class of Śramaṇas, the orthodox Brāhmaṇs' chief challengers in the competition for religious authority.

Digambara and Śvetāmbara tradition agree that by the time of his death Mahāvīra led a contingent of monks, nuns, laymen and lay-women, i.e., the fourfold order of Jain society. The numbers given by the *Kalpasūtra* belong to convention and tell us little about the size of Jain communities as they might actually have existed on the ground. It is, however, noteworthy that nuns outnumbered monks. It is sometimes said that this indicates that the current situation in which nuns greatly outnumber monks has always been the case. This is possibly true, but we have no evidence one way or the other.

From the standpoint of the Jains' own historiography, the absolutely crucial issue is the line of disciplic succession that connects Mahāvīra with subsequent generations of Jain mendicants. Such a line of succession can be considered the spiritual equivalent of a clan or lineage, and just as family genealogies do not always reflect literal historical truth – which is frequently unknowable in any case – so, too, spiritual genealogies often reflect the concerns and desires of later periods as opposed to the actual 'facts' of the past. In this sense, such genealogies serve as charter narratives that invest current realities with the legitimacy of antiquity. In the case of Jainism, the putative existence of a long line of connection to Mahāvīra is the decisive factor in the legitimisation of currently existing mendicant orders and, by association, the authenticity of the practice of their lay followings. I say mostly because there was at least one Śvetāmbara mendicant lineage (the Upkeśa Gaccha, apparently now extinct) that traced its spiritual descent to Pārśva, and there could well have been others (see Hoernle, 1890, for the lineage's disciplic genealogy).

Jain tradition maintains that at the time of Mahāvīra's death there were eleven *gaṇadhara*s (for details, see P. S. Jaini, 1979, pp. 43–6). Tradition further holds that nine of the *gaṇadhara*s became *kevalin*s (attained omniscience) during his lifetime. With this, their proselytising and preaching ceased, because only a Tīrthaṅkara teaches while in the enlightened state. The exceptions were Indrabhūti Gautama and another *gaṇadhara* named Sudharma. Some of the nine *kevalin*s died before Mahāvīra's death; the rest did so within the year of his death.

The reason Indrabhūti Gautama had failed to achieve omniscience prior to Mahāvīra's death was the spiritual flaw of his deep attachment to Mahāvīra. He did, however, attain omniscience in the immediate aftermath of Mahāvīra's death, and lived on as a *kevalin* for twelve years more, but his omniscience had brought his teaching and leadership role to an end. As a result, it was left to Sudharma to transmit the scriptures, of which he had complete knowledge. He imparted all that he knew to a disciple of his own named Jambū. Having transmitted the scriptures, Sudharma himself attained omniscience in the thirteenth year after Mahāvīra's death and died within a year of that. With Sudharma's death, all the *gaṇadhara*s were gone.

Jambū, in his turn, taught for eight years, and attained omniscience and liberation sixty-four years after Mahāvīra's death. He was the last person to attain liberation in our declining epoch and corner of the cosmos. Before his attainment of omniscience, Jambū passed the scriptures to the first of a line of four successors. The last and fourth of these, and thus the seventh successor within the spiritual lineage established by Mahāvīra, was Bhadrabāhu, the putative author of the *Kalpasūtra*. He is said to have been the last person to know all of the now-extinct texts known as the Pūrvas.

Early diffusion

It seems likely that, in the earliest days after Mahāvīra's departure, Jain teachings were spread by mendicants affiliated with proselytising mendicant groups of some sort (called *gaṇa*s in the texts). We know little about how these groups were organised, but it is virtually certain that central to their structure was, as we have already noted, the discipolic equivalent of unilineal descent in which affiliation was not established by birth but by initiation into a group tracing spiritual descent back to an original source, presumably to Mahāvīra himself via a chain of intermediary figures.

In social terms, Jainism undoubtedly found its richest soil for lay-proselytisation among the urban classes, particularly the traders of the day. It appears to have moved outward from the region roughly corresponding to modern Bihar in two directions: westward to Mathura (supposedly once visited by Mahāvīra) and beyond; and eastward into Bengal and then down the subcontinent's eastern coast through Odisha and into the South. In fact, there have been relatively few Jains in Bihar in later periods up to the present, but by roughly the third century CE Jainism had established itself throughout most of the Indian subcontinent – North and South.

The result of this bifurcated exodus was, in very general terms, the evolution of a significant difference between the religious cultures of Jains in the North and South that is manifested most dramatically in the split between the Śvetāmbaras (North and West) and Digambaras (South, but also present in the North).

The Śvetāmbara/Digambara divide

On the most basic points of doctrine there is little difference between the Śvetāmbaras and Digambaras, but the dispute that divides them – the question of male mendicants' clothing – is deeply felt. To the Digambaras, nudity is the essential precondition of true mendicant status. Both branches agree that Mahāvīra was nude, but the Digambaras say that he gave up clothing immediately, whereas the Śvetāmbaras maintain that he wore the cloth for a year and a month, after which it was accidently pulled off by a thorn as he passed by.

But in addition to the key disagreement about mendicants' dress, there are many other differences between the two branches. Śvetāmbara

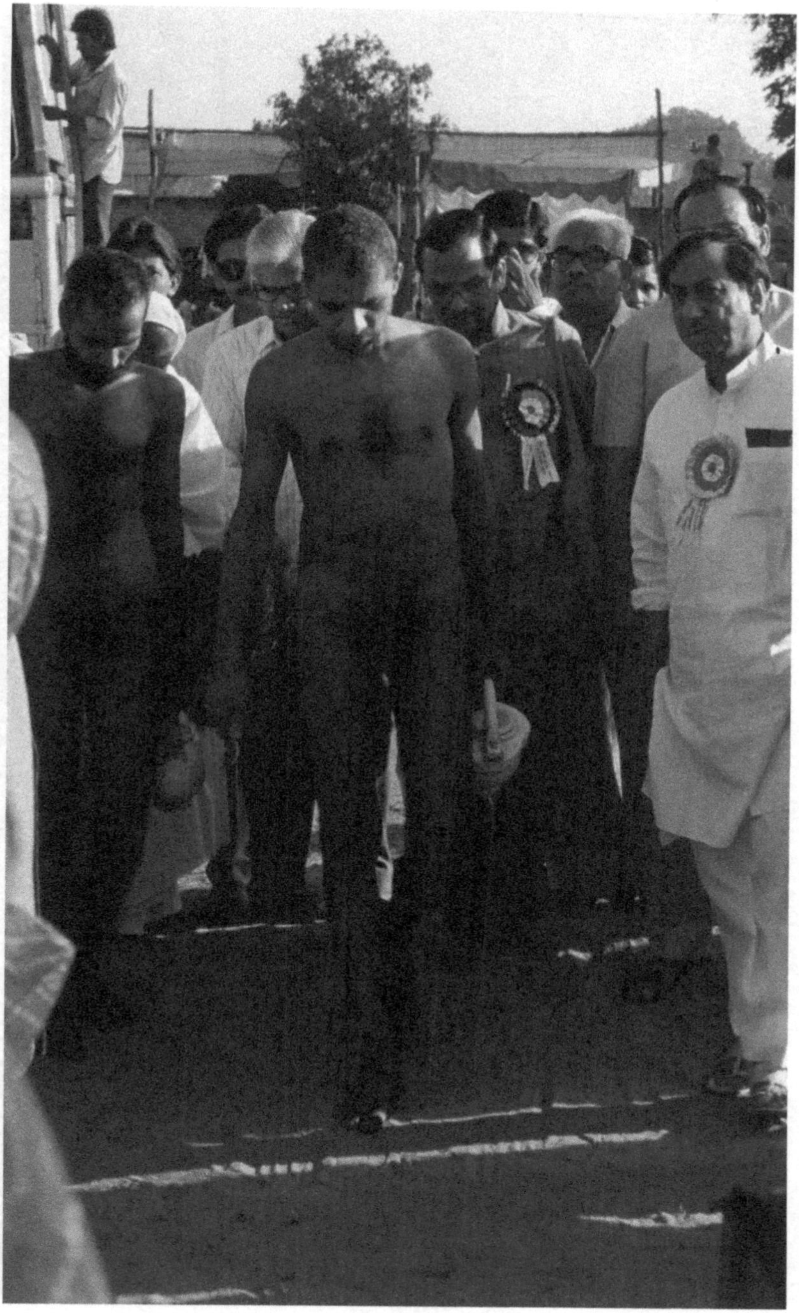

2.1 Two Digambara monks arriving at a ceremony.

2.2 A Śvetāmbara monk.

mendicants accept food-alms in bowls that they take back to their places of temporary residence for consumption; Digambara monks carry no bowl and receive their food in cupped hands while standing. Śvetāmbara mendicants can eat more than once a day, but Digambara mendicants can eat only once. While Śvetāmbaras and Digambaras have similar annual cycles of festivals and observances, the datings are different. The Śvetāmbaras believe that the Tīrthaṅkaras have human biological needs and therefore eat and excrete, but the Digambaras claim they do not (on which see especially Dundas, 1985). There are various other points of difference, large and small.

One difference, however, merits separate discussion. It has to do with the religious status of women (on which issue, see especially P. S. Jaini, 1991). The Digambaras believe that liberation cannot be achieved in a female body. This is mainly because a woman cannot practise nudity (essentially a cultural given); thus, she cannot be a true mendicant, and only a mendicant can achieve liberation. There are other reasons as well, but it should be borne in mind that a woman can certainly achieve liberation if she transmigrates to the body of a man. For their part, the Śvetāmbaras maintain that women can and do achieve liberation and that one of the twenty-four Tīrthaṅkaras of our current declining epoch, number fourteen whose name was Malli, was actually a woman. They also maintain that Marudevī, mother of Ṛṣabha, the first Tīrthaṅkara of our current declining epoch, was the first individual to achieve liberation in our epoch, which she gained even before her son.

The Digambaras and the Śvetāmbaras also have widely different views of the causes of their split. A Digambara version focuses on a twelve-year famine that is said to have occurred in North India *c.*360 BCE (P. S. Jaini, 1979, p. 5; K. C. Jain, 2010, Vol. 2, p. 728). They claim that Bhadrabāhu, the last of Mahāvīra's successors to have mastered all of the Pūrvas and author of the *Kalpasūtra*, led a group of monks into South India in order to escape the famine. With them was Chandragupta Maurya, first of the Mauryan emperors, who had become a Jain monk. Unable to return to the North before the end of the famine, Bhadrabāhu ultimately died by means of the ritual of self-starvation at Shravana Belagola (in what is today the Indian state of Karnataka). When the other monks returned to the North, they were met by a couple of ugly surprises. The first was that the northerners had adopted a spurious sacred canon. Worse yet, the northerners had backslid into the wearing of clothing. The naked monks completely rejected the northerners' canon and condemned their adoption of clothing.

According to a Śvetāmbara version (Dundas, 2002, p. 46; K. C. Jain, 2010, Vol. 2, p. 727), about 600 years after Mahāvīra's death a renegade (and self-initiated) Śvetāmbara monk by the name of Śivabhūti, having heard that Mahāvīra and his followers had abandoned clothing and arrogantly thinking himself to be at the same spiritual level, abandoned clothing himself. After his expulsion from the Śvetāmbara fold, he and his disciples established the Digambara branch.

The most likely reason for the split, however, is regional isolation and the amplification of difference over time. Mahāvīra and his disciples were almost certainly nude, but the earliest texts seem to suggest that wearing or not wearing clothes was seen as an option for mendicants in Jainism's early centuries (Dundas, 2002, pp. 47–8). In the medieval period, there existed a mendicant order in South India occupying an intermediate position. Known as the Yāpanīyas, they were normally nude but donned clothing of some sort while in public (Wiley, 2004, pp. 238–9). In any case, the differences between the two branches appear to have begun to crystallise around the fifth century CE and then to have widened over time into the chasm that we see today. The South ultimately became the heartland of the Digambaras; there they flourished and are found today mainly in Karnataka and southern Maharashtra. The Śvetāmbaras became predominant the North and West, particularly in Gujarat, Rajasthan and Punjab, but it should be borne in mind that there are also large numbers of Digambaras in North and Central India.

Texts and the sectarian divide

The *āgama*s (canonical literature) of the Jains cannot be discussed outside the context of the split between the Digambaras and Śvetāmbaras, because the matters are deeply intertwined. (A more complete discussion of the entire corpus can be found in P. S. Jaini, 1979, pp. 47–88.) On one point, however, the two branches agree; both maintain that the most ancient Jain scriptures have been lost. These texts are said to have consisted of fourteen bodies of material known as Pūrvas (old or previous) that are said to date from Parśva's time. Since Śvetāmbara and Digambara traditions generally agree on the content of these now-extinct materials, they must have existed in some form before the Śvetāmbara/Digambara split. About the subsequent history of Jain scripture, however, the disagreement between the two branches is deep and nearly total.

Let us look at the Śvetāmbara contentions first. According to the Śvetāmbaras, most of the Pūrva teachings were included in the twelfth section of the Aṅga texts (below), which is known as the *Dṛṣṭivāda* (disputation of views), but that the twelfth Aṅga (i.e., the *Dṛṣṭivāda*) has itself been lost. But they further maintain that there remained a body of scripture (not, of course, the Pūrvas) that they were able to authenticate at a council held at Pataliputra around 160 years after Mahāvīra's death, and again at a second council held bilocationally (at Mathura and Valabhī) around 800 years after his death, and yet again at a third council held at Valabhī in the fifth century CE, at which it is said that the canon was reduced to an authoritative version in writing. Although these occasions are usually called 'councils' in English, Dundas (2002, p. 71) points out that the term actually used to denote them is *vācanā* (recitation). The idea seems to have been to stabilise and legitimise an otherwise unstable body of orally transmitted material by means of its recitation by particularly learned monks.

This Śvetāmbara canon, written in a Prakrit called Ardhamāgadhī (in which Mahāvīra is said to have taught), is conventionally held by image-worshipping Śvetāmbara Jains to consist of forty-five texts that they regard as representing a significant portion of Mahāvīra's original teachings. (The two non-image-worshipping Śvetāmbara sects – the Sthānakavāsīs and Terāpanthīs – do not accept all forty-five as valid.) Although it is impossible to date these materials accurately, there is little doubt that some of their elements do indeed date from a time close to that of Mahāvīra himself.

The subdivisions of the materials are six. The first and most important are the eleven Aṅgas (limbs). As just noted, there were originally twelve, but the twelfth Aṅga (the *Dṛṣṭivāda*) is regarded as lost. The eleven existing Aṅgas contain a huge variety of material covering doctrine, monastic discipline, ascetic practice, Mahāvīra's life, stories of exemplary piety and much more. The Aṅgas possess a special importance and stand apart from the other four groupings because they were transmitted by the *gaṇadhara*s themselves, whereas the other lesser texts originated at a later period from learned mendicants known as *sthavira*s (elders), of which the first was Jambū.

Subordinate to the Aṅgas are twelve bodies of material known as Upāṅgas (subsidiary to the Aṅgas). Their content is, like that of the Aṅgas, highly variegated and includes points of doctrine, the birth of gods and hell-beings, forms of life, matter and soul, the nature of the sun, moon, an account of Jambū Dvīpa (the central island of the terrestrial disc; see Chapter 6), the biography of Ṛṣabha, accounts of exemplary lay piety, conversion stories and more.

The third grouping consists of six texts (there were originally seven, but one is now lost) known as Chedasūtras that are mainly concerned with issues of monastic discipline. Our *Kalpasūtra* belongs to the eighth chapter of the first of these texts. Fourth are the Mūlasūtras (root *sūtra*s), four in number, which probably bear their collective name because they are supposed to be learned by mendicants at the outset of their careers. They deal mainly with matters of monastic life and discipline. Fifth are the Prakīrṇakasūtras (miscellaneous *sūtra*s), consisting of ten texts on a range of topics from the proper ritualisation of death by self-starvation to praise of the Tīrthaṅkaras. Sixth and last are two texts called the Cūlikāsūtras (appended *sūtra*s), dealing mainly with epistemology and hermeneutics.

Believing that almost all the original canon was lost, the Digambaras reject the entire Śvetāmbara canon as bogus. The Digambaras, however, also maintain that the twelfth *Aṅga* was not entirely lost, and that in the second century CE a Digambar monk named Dharasena, thirty-third successor to Indrabhūti Gautama, orally taught an incomplete version of it to two of his disciples (P. S. Jaini, 1979, pp. 50–2). They in turn rendered it in writing – in a Prakrit with some Sanskrit – as the Ṣaṭkhaṇḍāgama (scripture in six parts), which may well be the first written scripture of the Jains. Somewhat later, a similar text entitled *Kaṣāyaprābhṛta* (treatise on the passions) was produced by a Digambara monk named Guṇabhadra. The Digambaras regard these two texts as the only surviving fragments of the original Jain canon, and in this respect their canonical literature, properly speaking, is quite small.

However, learned monks in both Śvetāmbara and Digambara communities produced a prodigious post-canonical literature that has acquired semi-canonical status, and these two textual traditions together constitute one of the glories of Indian religious and philosophical literature. A survey of these materials is completely beyond the scope of this book, but some high points should at least be mentioned.

Regarded as authoritative by both Digambaras and Śvetāmbaras is the *Tattvārtha Sūtra* (J. L. Jaini, 1974b; Tatia, 1994), a Sanskrit work written in the fourth or fifth century CE by Umāsvāti (also known as Umāsvāmī, mainly in Digambara tradition), in which we find an extraordinary range of Jain belief and doctrine presented in a highly systematised form. When we turn to cosmography in Chapter 6 we shall draw extensively from this work. Of particular note on the Śvetāmbara side are two scholar-monks of extraordinary distinction: Haribhadra and Hemacandra. Haribhadra was a seventh- or eighth-century figure (his dates are contested) who produced a vast amount of writing on virtually every branch of learning. Hemacandra, belonging to the twelfth century, was not only a great poet but also played a key role in the Gujarati politics of his day because of his close relationship with King Kumārapāla. On the Digambara side (but also admired by many Śvetāmbaras) is the second- or third-century monk-scholar Kundakunda. His writings are considered foundational to Digambara religious culture, and he is particularly noted for his mystical bent; he regarded omniscience as deep and complete self-knowledge or self-realisation, experienced by means of inward contemplation (see, e.g., J. L. Jaini, 1974a).

The Digambara South

Although Digambara Jains were and are today present in the North, the South Indian Jain world was from the start and remains today dominated by Digambaras. This is so to the extent that a history of southern Jainism is virtually the history of Digambara Jainism. According to legend, Samprati, a grandson of the emperor Ashoka, was responsible for Jain proselytising in the South. Supposedly, he first sent sham Jain monks to the South in order to teach the locals how to feed and support Jain mendicants, and then followed up with the real thing. Dundas (2002, p. 116) plausibly interprets this as a mythologised recollection of the penetration of the South by groups of mendicants 'who acted as transmitters of a northern, prestigious culture, of which Jainism was a part'.

Evidence for the early influence of Jainism in what is now the state of Odisha is provided by a famous inscription at the Hāthīgamphā cave. The cave's name means 'elephant cave', and it is located near the present-day city of Bhubaneswar. The inscription, which is poorly preserved and difficult to interpret and date, was created at the command of Khāravela, king of Kalinga, *c.*150 BCE to celebrate the military achievements of his reign. Khāravela was probably a Jain or Jain sympathiser because the inscription refers to an image of a Tīrthaṅkara that had earlier been looted from Kalinga and that he returned to the kingdom (Cort, 2010b, pp. 39–41).

It is possible also that Jain mendicants and lay communities existed in Tamil country as early as the third century BCE, even before the diffusion of Vedic traditions from the North (K. C. Jain, 2010, Vol. 2, p. 440). But whatever the dating of its arrival, Jainism, in its Digambara (or proto-Digambara) version, then went on to an extraordinarily successful career in the South India. From roughly the fourth to the tenth centuries, Jains appear to have been extremely effective at attracting patronage from South Indian warriors and kings, especially in what is now Karnataka. Jainism bore the prestige of the North, thus possibly providing a means for martial elites to differentiate themselves from their rustic subjects (Dundas, 2002, pp. 118–20).

In view of the Jain commitment to an ethic of non-harm, this apparent affinity with warriors might seem incongruous, but it should be remembered that Jains venerate Jinas, spiritual 'victors' or 'conquerors', who turn valour away from the physical battleground and into the fight against the body and its desires and aversions. This is a symbolism that could easily resonate with the sensibilities of a martial elite. At the same time, Jains never truly deviated from the view that worldly conquest was quite inferior to the inner conquest of desire and aversion, and that kingship was only worthy of 'partial admiration' (Dundas, 1991, pp. 180–4).

A crucial development among Digambaras was the emergence in the early thirteenth century of a class of non-peripatetic and clothing-wearing religious authorities known as *bhaṭṭāraka*s whose disciplic lineages appear to have been offshoots of those of the naked mendicants (Dundas, 2002, pp. 123–5; Flügel, 2006, p. 339). Their evolution as a class was probably an inevitable result of the accumulation of donated property that had to be managed by someone, but Muslim rule and Muslim disapproval of public nudity might also have played a role. In the end, the *bhaṭṭāraka*s exercised enormous authority over the Digambara populations living in the hinterlands of their monastic seats; in effect, they became the theocratic rulers of Digambara caste communities. During their heyday the Digambara institution of naked monks went into severe decline, not fully to recover until the twentieth century.

Ultimately, the *bhaṭṭāraka*s of the northern Digambaras were seriously challenged by a Digambara reformist movement that came to be known as the Terāpanth (Path of Thirteen, as opposed to the Bīspanth, the Path of Twenty who supported the *bhaṭṭāraka*s) and that emerged in the seventeenth century in the Rajasthani town of Sanganer (Cort, 2002b). This movement brought about the end of most of the *bhaṭṭāraka* seats in North India, but not the *bhaṭṭāraka* seats in the South (Flügel, 2006, p. 346). Lay reformers in the South began to introduce a restructuring of the management of the *bhaṭṭāraka* seats in the late nineteenth century, but the most crucial development was the revival of the Digambara tradition of mendicant nudity in the twentieth century. This occurred under the aegis of three distinguished Digambara monks: Acharya Shantisagar (1873–1955), Acharya Shantisagar (Chani) (1888–1944) and Acharya Adisagar (1868–1943) (Wiley, 2004, pp. 25, 192). The re-emergence of naked monks inevitably devalued the religious authority of the *bhaṭṭāraka*s, but, even so, they remained a key element in the pattern of Southern Digambara Jainism to the present-day.

While Digambara Jainism has remained well entrenched in the South, it underwent a serious decline after reaching its apogee in roughly the sixth century. This was a result of the bitter hostility of the Hindu sects of the region from roughly the seventh century onward. Usually characterised as a 'Hindu revival', this development was at least in part a result of sectarian competition for royal patronage. Famously, the hostility was expressed in the compositions of Śaiva and Vaiṣṇava devotional poets; indeed, it has been plausibly suggested that the Jains were the principal 'other' against which the Śaiva devotional religious culture – itself central to Tamil cultural identity – was forged (Peterson, 1998). While persecution of the Jains led ultimately to their marginalisation in the South, it must also be recognised that even during those dark days Jains and Hindus were engaged in creative interaction, with the result that Jainism left a distinct mark on some South Indian Hindu traditions, most notably Śaiva Siddhānta (Davis, 1998).

The mostly Śvetāmbara North

According to historian Smita Sahgal (1994), the westward movement of Jainism occurred mainly in the post-Mauryan period, *c.*200 BCE – 300 CE. She argues that in this region – and in contrast to India's East and South – Jains had little royal patronage, which was mainly directed toward Buddhism and Brāhmaṇical orthodoxy. Rather, the context in which Jainism gained a foothold was the economic development and expansion of trade that took place there, especially under the Kushanas, whose empire drew Central Asia and North India into the same trading zone. An additional factor favouring Jainism was the religious tolerance of the Kushanas themselves. These trends favoured the urban Vaiśya community, especially the traders among them, who rose in wealth and relative social status. They were the social base of Jainism in this region, which must have been propagated by proselytising monks.

Jainism took root in the city of Mathura, an important trading centre of the day, in roughly the second century BCE, and inscriptional evidence suggests strong lay support from the trading community (Chatterjee, 1978, p. 67; see also Folkert, 1989). This city ultimately became a regional Jain epicentre. In those days, as now, Vaiṣṇavism and the Kṛṣṇa ('Krishna') cult were especially strong in Mathura, and the Jain assertion that Kṛṣṇa was actually a Jain layman (and a kinsman of Nemi, the twenty-third Tīrthaṅkara) may reflect a conversion strategy dating from that time and place (Dundas, 2003, p. 136). By the fourth and fifth centuries CE, Jainism apparently waned in the Mathura area and its centre of gravity in India's North moved towards the region that is now Rajasthan and Gujarat. Gujarat, where there is evidence of its presence as early as the second century CE, ultimately became the stronghold of Śvetāmbara Jainism.

The subsequent development of Śvetāmbara Jainism is somewhat obscure, although some light is thrown on the period by hagiographies of great Śvetāmbara teachers, proselytisers and miracle-workers. Undoubtedly, the most notable figure of Śvetāmbara history in India's West was Hemacandra (Dundas, 2002, pp. 134–6; K. C. Jain, 2010, Vol. 2, pp. 598–9). He was born in 1089 (d.1172) to a family belonging to a trading caste and was initiated as a monk in boyhood. He was precocious and able, and rapidly rose to the rank of *ācārya* (mendicant leader) within his mendicant community. He is remembered best for his extraordinary learning, prodigious literary output and, above all, his close relationship with King Kumārapāla (r.1143–1172), one of Gujarat's most celebrated rulers. Supposedly, Hemacandra assisted Kumārapāla, then not yet king, in escaping an assassination attempt by his uncle. Whether this is true or not, their relationship was indeed close, and Hemachandra is generally credited in converting him from the Śaiva sect of Hinduism and persuading him to rule in accordance with Jain teachings.

By the eleventh century, the Śvetāmbaras of northern India began to use the term *gaccha* to refer to their disciplic mendicant lineages (Dundas, 2002, pp. 138–45), and a number of these lineages emerged on to the historical stage, each led by an *ācārya* (bearing the title *sūri*). Although these were mendicant groups, each was linked with a lay community from which it derived material support. Thus, and in consonance with the Jain ideal of the fourfold social order, the *gaccha* was at the apex (at least in theory) of a community of monks, nuns, laymen and laywomen. There was probably much rivalry between these lineages for lay followers.

A major issue among Śvetāmbara Jains of this period was the question of the status of non-peripatetic monks. By the early centuries of the Common Era, a class of *caityavāsī*s (temple- or monastery-dwelling monks) had emerged among the Śvetāmbara Jains of North India, a pattern that duplicates a similar development (the *bhaṭṭāraka*s) among Digambaras. The practice lasted for centuries in the Śvetāmbara world and was defended by those who engaged in it, but it is hard not to view it as a form of backsliding when seen against the ideal of the wandering monk. Challenges to the temple-dwellers came into historical visibility in roughly the eleventh century, but the problem of domesticated monks proved durable and played an important role in the formation of mendicant lineages.

Of the *gaccha*s that emerged in medieval times, the best known are two: the Kharatara Gaccha and the Tapā Gaccha. (There were and are Śvetāmbara *gaccha*s other than these two; these will be discussed in Chapter 4.) The Kharatara Gaccha originated as a reform movement aimed at the temple-dwelling monks (Babb, 1996, pp. 114–15; Dundas, 2002, pp. 140–2). Its founder was a monk named Vardhamānasūri who, although initiated as a temple-dweller, left his original mendicant community and preached against the temple-dwellers until his death in 1031. His most distinguished disciple was a deeply learned monk of Brāhmaṇ origin named Jineśvarasūri, who played a key role in enhancing the lineage's prestige and influence. According to one version of the story, Vardhamānasūri had gone to Patan, a hotbed of the temple-dwellers, in order to confront them, and was challenged there by the temple-dwellers to a debate. The debate took place in 1024 CE in a Pārśva temple in the presence of the king. Jineśvarasūri carried the burden of the debate with great distinction and was totally victorious. The admiring king then characterised Jineśvarasūri as *khartar* (fierce) in debate, which was how the *gaccha* got its name. The Tapā Gaccha was established in the thirteenth century by one Jagaccandrasūri, who had left his previous lineage because of what he considered laxity in ascetic practice (Dundas, 2007, pp. 34–52). The intensity of his own ascetic practice gave rise to the *gaccha*'s name, *tapā* being a term for asceticism.

The Tapā Gaccha and Kharatara Gaccha maintained a rivalry for many centuries that even endures today. In recent times, however, the Tapā Gaccha has become the more influential and powerful of the two by far, especially in the Śvetāmbara heartland of Gujarat, and has been much more successful in recruiting monks. The Kharatara Gaccha remains more influential in some areas of Rajasthan, and has left an impressive legacy in the origin narratives of many clans belonging to the Osvāl caste, who believe their ancestors to have been converted by Kharatara Gaccha *ācārya*s.

A slow crisis for the Jains in the North and West was the coming of Muslims and the advent of a series of Muslim polities from the twelfth century onward, culminating in the emergence of the mighty

Mughal Empire in the sixteenth century. A bad start was the destruction of Valabhī, the site of the fifth century Council of Valabhī, by Turks in CE 782. And while relations between Jains and their Muslim rulers blew hot and cold, the early centuries of Muslim rule saw the widespread despoliation of Jain temples and places of pilgrimage, of which the destruction of the temples at Mount Shatrunjaya in 1313 is perhaps the most important instance. Nonetheless, Muslim rulers for the most part came to terms with Jain communities, a necessity given their economic importance.

The Mughal emperor Akbar (1542–1605) was a man of great intellectual gifts and generosity of spirit. He had a deep interest in all religions, including Jainism, and he appears to have had an especially close relationship with a Tapā Gaccha *ācārya* by the name of Hīravijayasūri, under whose influence he banned the slaughter of animals during the annual period of Paryuṣaṇa (Dundas, 2002, pp. 146–7; K. C. Jain, 2010, Vol. 3, pp. 1145–9). He is also said to have invited the Kharatara Gaccha *ācārya* Jincandrasūri (one of the Dādāgurus; see Chapter 4) to court and conferred on him the title *yugpradhān* (leader of the age). Akbar's successor, Jahangir, apparently took a more ambivalent view of the Jains; on the one hand, he appointed a Jain tutors for his son, but, on the other, he apparently tried to ban Jain monks from the empire.

Both the Kharatara Gaccha and the Tapā Gaccha belong to a tradition that came to be known as Mandirmārgī (temple-going) or Mūrtipūjaka (image-worshipping). They, along with other Śvetāmbara *gaccha*s emerging in medieval times, supported the legitimacy of the worship of images in temples, as did all but one of the subgroups among the Digambaras (the small Tāraṇ Svāmī Panth; see Cort, 2006a). For the Śvetāmbaras, however, the issue of image-worship became momentous indeed, and the map of the Śvetāmbara world, as it exists today, was decisively shaped by reformers who vehemently opposed image-worship as a practice alien to Mahāvīra's actual teachings and nowhere sanctioned by valid scripture.

The critical figure is this context was a Gujarati named Loṅkā Śāh who lived in the fifteenth century (and on whom see Cort, 2010b, pp. 226–34). His historical persona is deeply embedded in hagiographic mythology and little is actually known of him. The basic Sthānakavāsī account (that is, an account within the tradition that grew out of the reforms he initiated) represents him as a lay Jain and scribe to whom a lax monk of Ahmedabad gave some manuscripts of important texts for recopying. In the course of his copying he became aware of the very great difference between the monastic discipline described in the texts and the actual behaviour of the monks of his place and time. He became convinced of the basic illegitimacy of image-worship and certain other features of practice that he believed had departed from Mahāvīra's strict standards.

Although Loṅkā Śāh was probably never initiated into an existing disciplic lineage, he apparently did lead a mendicant life of a sort, and his followers became the nucleus of a new Loṅkā Gaccha, unanchored in any existing lineage. The *gaccha* survived his death but appears to have been rapidly overtaken by compromise and backsliding, for it seems that the Loṅkā Gaccha had itself begun installing images in temples before a century had passed (Dundas, 2002, p. 251). Only a small remnant of the Loṅkā Gaccha survives today.

Although the Loṅkā Gaccha met a dismal fate, a more lasting reform tradition emerged from it in the seventeenth century. Two laymen, each of whom rejected image-worship and sought a return to the ideal of severe monastic discipline, left the Loṅkā Gaccha, and each established his own monastic community. This was the origin of the cluster of mendicant lineages bearing the name Sthānakavāsī today. The term means dwellers in *sthānaka*s (places), by which is meant places other than the community halls set aside for the sojourns of mendicants by image-worshipping Śvetāmbaras and often associated with temples. An innovation introduced by the Sthānakavāsīs (and retained by the offshoot Śvetāmbara Terāpanthīs) was the practice of the permanent wearing of the *muhpattī*, a cloth shield worn across the mouth designed to keep the wearer's hot breath from harming living things in the air. It is an indication of the notable depth of the Sthānakavāsīs' commitment to the ethic of non-harm.

Although the Sthānakavāsīs experienced setbacks in the nineteenth century, they resurged in the twentieth century and remain a major presence in the Śvetāmbara world today. In contrast to the Mandirmārgīs, among whom viable mendicant lineages are currently very few, the Sthānakavāsīs are a congeries of more than two dozen

2.3 A group of Śvetāmbara Terāpanthī nuns. Note their mouth shields.

mendicant groups. The religious life of the Sthānakavāsīs has a markedly different orientation from that of the Mandirmārgīs, precisely because of the absence of temples. The lack of temple ritual greatly accentuates the centrality of mendicants to lay religious practice, and the absence of temple construction and renovation as a destination for donations has gone hand-in-hand with the Sthānakavāsīs' well-known propensity for a great range of charitable activities.

The Sthānakavāsī movement, in its turn, gave rise to yet another aniconic reformist movement in the eighteenth century. Known as the Terāpanth (and not to be confused with the Digambara movement bearing the same name), this branch was founded by a Sthānakavāsī mendicant named Bhikṣu, by all accounts a stern and puritanical figure (Dundas, 2002, pp. 254–62; Wiley, 2004, pp. 57–8, 216–17). He was born in 1726 (d.1803) in the region known as Marwar (now part of Rajasthan) to a family belonging to the Osvāl caste, a trading caste prominent in the region, and Rajasthan has always been the heartland of the movement he founded. He was initiated into a Sthānakavāsī mendicant lineage in 1751. After eight years, and having detected what he saw as serious backsliding among his fellow monks, he and six other monks broke away from this lineage and formed their own mendicant community. Its name, Terāpanth, can be rendered either as 'path of thirteen' or 'your path', the latter presumably meaning Mahāvīra's path. A distinctive feature of Bhikṣu's teachings – often criticised by other Jains – was that charitable activities and gifting benefiting non-mendicants are purely worldly activities and devoid of spiritual significance.

After various tribulations, the Terāpanth became the extremely successful branch of Śvetāmbara Jainism that flourishes today. It is known for its highly centralised structure under the authority of a single *ācārya* (see Flügel, 1995–6, on the Terāpanth's organisation). In respect of its centralisation, the Terāpanth may be considered the exact opposite of the highly fragmented Sthānakavāsī movement.

Jainism abroad

Jains had been involved in India's overseas trade for centuries, but it was not until the late nineteenth century that Jains actually began to create permanent overseas communities. The first chapter of this story was the migration, under the umbrella of the British Empire, of significant numbers of Gujarati Jains (both image-worshipping Śvetāmbaras and Sthānakavāsīs) to East Africa in search of entrepreneurial opportunities. The Jains (and Hindu traders also) did well in Africa, but their affluence and success were resented by locals, and when Kenya, Tanzania and Uganda became independent in the 1960s life became difficult for Indians in these countries; indeed, they were simply expelled from Uganda in 1972.

The result was a second migration that took many Gujarati Jains to the UK and some to the USA and Canada. In the meantime, US immigration laws were liberalised in 1965, which enabled significant numbers of Jains and other Indians, both from Africa and India itself, to settle there. The actual numbers involved in these migrations are extremely uncertain. By one estimate (Wiley, 2004, p. 19) roughly 100,000 to 150,000 Jains currently live abroad, and of these about 50,000 have settled in the USA and Canada, 30,000 in the UK and Europe, 20,000 in Africa and 5,000 in Asia outside of India.

In the religious sphere, the principal difficulty faced by Jains abroad has been the absence of mendicants owing to the prohibition of their travel in vehicles (although Digambara *bhaṭṭāraka*s are not prohibited from such travel). The problem is only partly one of doctrinal expertise; lay experts can communicate Jain teachings, as can Internet websites, and in fact there is currently a strong online presence of websites providing information about Jain teachings and practices and aimed at expatriate Jains. More to the point is that Jainism, as currently practised by the majority of Jains in India, involves frequent interaction with mendicants who teach, encourage, inspire, sometimes scold and in general exercise a religious authority that is conferred by world renunciation and formal initiation and is unavailable to lay Jains, no matter how learned. In both the UK and North America, Jains have to some extent filled the gap with a rich associational life in which sectarian differences have receded in importance. Indeed, the absence of mendicants has probably facilitated this, because the exhortations of mendicants are factors in the maintenance of sectarian boundaries in India.

At the same time, the barrier to mendicant travel abroad has been challenged and shows signs of at least some weakening. The ice was broken by two remarkable figures. In 1970, a Śvetāmbara Mandirmārgī monk by the name of Chandraprabhasagar (b.1922) travelled to a religious conference in Geneva, and in 1971 journeyed to New York where he founded the Jain Meditation International Center. Now known as Chitrabhanu, he left the mendicant life, married and has continued to advocate reform of Jain traditions along rationalist and modernistic lines (Wiley, 2004, p. 67). In 1975, a progressive Sthānakavāsī monk named Sushil Kumar (1926–1994) embarked on an international tour without renouncing his monastic vows, an event that was regarded as quite scandalous at the time. A tireless promoter of interreligious harmony and non-violent causes, his best-known achievement was the establishment of a temple-ashram complex called Siddhachalam in Blairstown, NJ in 1983 (Wiley, 2004, pp. 207–8). It is often said to be the first Jain place of pilgrimage outside India, and is an excellent example of an ecumenism typical of Jainism abroad, with images suitable for both Śvetāmbaras and Digambaras included in the temple.

On another front, a truly radical innovation came in 1980 from the Śvetāmbara Terāpanthīs. They were then led by an extraordinarily charismatic and visionary leader, Acharya Tulsi (1914–1997), who established a new kind of mendicant, known as Samaṇ (male) or Samaṇī (female), who are permitted to travel by vehicle. He did so precisely to serve overseas Jain communities, which these mendicants are indeed doing today.

But such instances as these notwithstanding, it is hard to imagine that Jainism abroad will ever possess a monastic core of the sort seen in India, and, as this book will show, a sustained and intimate relationship between mendicant and lay communities has been the very heart of Jain tradition from its earliest days until now. Thus, we may be confident that whatever Jainism becomes outside of India – and on this point the jury is still out – it will present a range of significant and interesting departures from its Indian paradigm (on this point, see Banks, 1992, esp. pp. 196–217; for dietary issues, see Vallely, 2004).

Chapter 3

Liberation's Roadmap

Jain life – including Jain religious life – is about many things, not just about renouncing the world. The quest for worldly well-being has a definite role in Jain belief and practice, a fact that accounts of Jainism have too often ignored. Still, we must not mince words. There is a perspective at the core of Jain life, that of the *mokṣa mārga* (path of liberation), that is uncompromising in the severity of its ascetic and ethical demands. This perspective, which takes the form of the soteriological doctrines presented in the Jains' sacred writings and promoted by mendicant communities, often fades into the background of daily life, and most Jains regard liberation as a very distant goal indeed. Still, it never ceases to exert its gravitational pull – sometimes weakly, sometimes strongly – on Jain religious culture. But what, we must ask, is meant by 'liberation'? Liberation *of* what and *from* what? And how is liberation to be achieved?

Jain teachings begin with the understanding – itself a premise beyond the reach of debate or doubt – that the death of the body is not the death of the soul, and that the soul is then reborn in a new body. As we learned in the preceding chapter, this is an understanding that Jainism shares with the generality of Indic religious traditions. This ceaseless passage from one body to another is called *saṃsāra*, a term that also denotes the world as such, for it is the arena in which we must make our way from birth to death to birth again, endlessly and without respite.

There are deep disagreements between Indic traditions about what exactly the soul or self is, but there is consensus on one point, which is that rebirth – however conceived – is not a good thing. Death inevitably awaits us at the end of life's journey, and, although life does indeed offer its pleasures and satisfactions, these are transient and offset by disappointment and loss, inevitably made final at death's door. Taken in pieces, therefore, one's transmigratory career in the world has its good points to be sure; but taken in its entirety, it is a concatenation of misfortune and a running calamity that anyone who takes an informed, long-term view will certainly hope to escape.

Different Indic traditions have various ways of construing this negative conceptual framework, and some – especially those focused on redeeming deities – promote a surprisingly positive and hopeful outlook. Much darker is the Jain perspective. Not only do Jain teachings take a negative view of worldly existence, but the inherently dismal character of our bondage is accentuated by an emphasis on infinities. The Jains reject the idea of a creator deity or indeed a creation of the world in any sense. Time has no beginning or end, and our journey through the world has already been going on from beginningless time, and has meandered its way through every corner of the vast cosmos. Indeed, by mathematical necessity the entire journey has already taken place infinite times. And more, our journey with its aggregate of sufferings will continue forever unless brought to a stop. That stop is liberation. This is the 'from what' of liberation.

But how is liberation to be achieved? On this point, Jain teachings encapsulate in a deceptively simple formula what is, in fact, a very complex intellectual and normative structure. The formula is that the path to liberation requires three things known as the *ratnatreya* (three jewels). They are *samyak darśana* (right faith), *samyak jñāna* (right knowledge) and *samyak cāritra* (right conduct). *Samyak darśana* refers to a deep confidence in the truth of Jain teachings, the foundation of which is the omniscience of the Tīrthaṅkaras. *Samyak darśana* is obviously foremost among the three, for without such faith knowledge cannot convince and a code of conduct cannot compel. *Samyak jñāna* is the possession and understanding of Jain teachings, and *samyak cāritra* is behaving in a manner consistent with those teachings. But the three jewels are merely a gateway into the complex soteriological system to which we now turn.

The soul

What is the nature of the entity that seeks liberation from the bondage of *saṃsāra*? The Jains denote it by the term *jīva*, a term that carries the basic meaning of 'life'. One author (Dundas, 2002, p. 93) translates it into English as 'life-monad'. This rendering comes as close to the real meaning of *jīva* in English as any I have seen, but it is somewhat cumbersome. Most writers on the subject seem to prefer 'soul', and that is the term used in this book.

Jain teachings divide all of reality into two broad and opposed categories: *jīva* (soul) versus everything that is *ajīva* (non-soul). These two constituents of reality are utterly unlike, and in this respect a radical dualism lies at the very heart of Jain teachings. The *ajīva* category encompasses both attributes and constituents of material reality. These include motion, rest, space, time (although the status of time as an independent existent is debated) and atoms. This particular way of categorising the *ajīva* side of reality is supported by rather arcane arguments that need not detain us. The term 'atom', however, is important. This English term is a somewhat inexact translation of the word *pudgala*. The *pudgalas* are tiny particles, infinite in number, indivisible and without extension. In aggregate, they constitute matter, and in their varied combinations they form extended objects and substances that exhibit the physical characteristics of the world we experience through our senses. Matter is the prison of souls, for souls, except those in the liberated state, are embodied, and this embodiment is the source of unceasing rebirth.

Although the bodies that imprison souls are mortal and come and go, the soul itself is immortal; it was never created nor will it ever cease to be. Souls are infinite in number and they perpetually circulate through the cosmos from birth to death to birth again. It should be noted that this concept is radically different from that of the world-soul of the *advaita vedānta* (mentioned in the previous chapter). To the Jains there is no question of an identity between the self or soul of the individual and an ultimate unifying reality; there is no such unifying reality, and souls are absolutely real in their individuality and plurality. The Jain view is obviously also very different from the Buddha's doctrine of *anātman* (no-soul or no-self).

Although souls are embodied in a multitude of different ways, all souls are in essence identical because they bear exactly the same defining qualities: *caitanya* (consciousness), *sukha* (bliss) and *vīrya* (energy) (on which see P. S. Jaini, 1979, pp. 104–6). These qualities are always present in the soul, but when the soul is in an embodied state they cannot fully manifest themselves. By *sukha* is meant an extraordinary delight or satisfaction that arises from the soul fully knowing itself, and it is normally disfigured or clouded by the distractions attendant to embodiment. By *vīrya* is meant the basic force that makes possible perception, cognition and our engagement with things of the world. As is *sukha*, it is limited and misdirected by the conditions of embodiment. Most crucial, however, is *caitanya*, which is the soul's capacity to perceive and know the things of the world and itself. Consciousness is fully manifested as omniscience, but omniscience is normally occluded by embodiment; only in liberation or at the threshold of liberation does the soul realise its inherent capacity to know all things.

But having said that all souls are exactly alike, we must add one important qualification. Some souls possess *bhavyatva* – an innate capability to be liberated – while others do not (P. S. Jaini, 1977). This capability has only to do with liberation; souls without it can even be reborn as deities, but they cannot escape the grip of *saṃsāra*. It should be stressed that this is not an Indic version of the Calvinists' predestination. Possessing this quality does not in itself guarantee liberation, for even to embark on the path to liberation one must find oneself in a situation in which one's innate capacity can be awakened, such as having contact with Jain teachings, and following that path presents many extraordinary difficulties and hardships.

The soul has no innate physical size or shape because it is not a physical entity. It does, however, possess dimensionality, for in its varied embodiments it acquires the size and shape of the bodies it inhabits in the same manner – it is often said – that the light of a lamp will fill a room, regardless of its size or shape. When the body dies, the embodied soul transmigrates to a new body and does so in a mere instant and in a straight line. The nature of the soul's new habitation depends on how its previous life or lives were lived. The soul makes the journey inside a covering that is material but highly sublimated.

The reason for the necessity of a material vehicle is that a record of the past cannot be inscribed on the non-material soul.

The instantaneous and undetoured character of the soul's journey from old to new body is a distinctive feature of Jain teachings and an important point of contrast between Jainism and Hinduism. In Hindu tradition, a significant betwixt-and-between period begins at the moment of death and ends with gestation in a new body. During this period, the soul must be aided in its onward journey (lest it linger malevolently) by means of funerary rites conducted by Brāhmaṇ priests and other specialists (see Parry, 1994 for a particularly rich account of such rites). Paul Dundas (2002, p. 103) is undoubtedly right in his suggestion that the Jains' distinctive views of the soul's instantaneous transition served to distinguish the Jains from the Hindu communities around them by foreclosing the possibility of extended funeral rites, and in so doing also kept Brāhmaṇ priests, always to some extent adversaries of the Jains, out of the picture.

It has been seen that the embodied soul has already been transmigrating for an infinite period of time and will go on doing so for all of infinite time to come unless there is some kind of intervention. The soul, that is, is deeply entrenched in embodiment, and we must now ask what accounts for the persistence of the soul's embodiment and rebirth. The answer is supplied by an extremely important concept, that of *karma*.

Karmic bondage

The term *karma* is an important word throughout the family of Indic religions, but this is especially so in Jainism. As noted in Chapter 2, the term carries the primary meaning of 'action' and a derivative meaning relating to the consequences of action. *Karma* (action) generates *phala* (fruit), and the nature of the *phala* depends on the nature of the action. Good yields good, bad yields bad, and one must inevitably eat the fruits of one's actions – that is, experience their consequences – if not in this lifetime then another. The issue is not one of divine judgement, for there is no judge; it is simple cause and effect.

With all of the above the Jains agree, but they have also developed a highly distinctive understanding of *karma*, one that provides much of the foundation for Jain soteriology and that separates Jainism quite decisively from other Indic religions. According to Jain teachings, *karma* is actually a type of matter. It takes the form of *pudgalas* (atoms) that are ubiquitous, floating free in every corner of the cosmos, and all unliberated souls are in continuous interaction with them. Whenever we act, our actions attract this matter and bring about its *āsrava* (influx) and ultimately its adhesion to and binding of the soul (*bandha*; bondage).

In respect to the influx of *karma* (as differentiated from its adhesion), Jain teachings stress the 'doing' side of action as opposed to its motivation. Action as such attracts karmic matter, which is consistent with the physicality of the Jains' image of *karma*. But simply because karmic matter is drawn into influx by action does not in itself mean that it sticks and holds the soul in its binding grip. Here Jainism confronts a problem inherent in any radical dualism. If the two constituents of reality are totally unlike in principle (as is true of soul and non-soul), how can they interact with each other? Just so, if *karma* is a form of matter and the soul is non-corporeal, how is it possible for *karma* to adhere to the soul? This question is perhaps never fully set to rest, but Jain teachings maintain that this is where the intentions behind action come into the picture. Mental dispositions (such as false beliefs) and the related *kaṣāya* (passions) that motivate actions cause karmic matter to adhere to the soul. One metaphor likens the soul to a cloth that passion causes to be wet; the wetness results in the adhesion of *karma*, just as dust sticks to a wetted surface (P. S. Jaini, 1979, pp. 112–13).

Jain teachings further maintain that karmic accumulations defile or stain the soul, and in so doing impart a *leśyā* (colour, stain) to the soul depending on the motivations underlying its actions. The colouring thus indicates the soul's spiritual and moral development; dark colours point to a low level of achievement, and light colours, particularly white, show a high level. Furthermore, there are different types of karmic matter. Although karmic matter is without particular qualities while afloat in the environment, it acquires some of the character of the actions that caused its influx and adhesion when it adheres to the soul.

Jains have devoted a great deal of care and attention to classifying and analysing different types of *karma*, which is hardly surprising given the importance of *karma* as the crucial factor in the situation of all unliberated souls. The result is an entire intellectual system, unique to Jainism, based on the ways in which *karma* interacts with the soul (for details, see P. S. Jaini, 1979, pp. 115–27).

Jains distinguish eight separate types of *karma*, organised into two subcategories of four each. One group of four, the *ghātiyā* (harming) *karma*s, have an obstructive effect on the soul's attributes, i.e., its abilities to know and perceive properly, to feel its inherent bliss and to mobilise its inherent energies. Each of these categories is itself divided into a complex system of subcategories. The other group of four is called the *aghātiyā* (non-harming) *karma*s. These are *karma*s that actually carry the moral effects of our actions into the future and shape future embodiments. Depending on the type, they determine one's future proclivity to feel pain or pleasure, category of rebirth, longevity and the conduciveness to spiritual growth of one's future environment.

A given type of *karma* naturally falls away from the soul when it has produced its effect; it then floats free until caught in the karmic influx of some other soul. But in the process of rendering its effect, *karma* generates the influx of new *karma*, which then adheres to the unliberated soul. Thus develops a karmic feedback loop that is the snare that pulls us into worldly bondage. To seek liberation, therefore, must require a radical interference with this process. But how is this to be done?

Liberty

Mokṣa (liberation) is a condition that can be achieved only by the complete removal of *karma* from the soul, thus allowing the soul to realise its true nature, which in the embodied state is occluded by karmic coverings. Such a soul is *vītarāga*, entirely devoid of the passions of desire and aversion. But the physicality of *karma*, as Jains conceive it, has an important bearing on how such a condition is to be achieved. To the Jains, liberation cannot be attained by means of a special sort of self-awareness or introspective knowledge alone. Knowledge is indeed a condition of liberation, but not a sufficient condition. This is because the karmic deposits that occlude self-knowledge are physically real and cannot be thought away; they must be actually removed. For this there are two strategies. First, the influx and adhesion of *karma* must be reduced and finally eliminated (*saṃvara*). Second, the karmic deposits already present must be shed (*nirjarā*). Let us examine these steps one at a time.

The reduction of karmic influx and adhesion is partly a matter of regulating behaviour. One must avoid the sorts of actions that most encourage karmic influx. Violent actions, which in their very nature both arise from and nurture the worst of our passions, top of the list of such actions, and therefore the cessation of violence is a crucial prerequisite to progress on the path to liberation. This is one of the foundations of the Jains' strong commitment to *ahiṃsā*. It is not the only foundation, because *karuṇā* (compassion) for all forms of life, as exemplified by the Tīrthaṅkaras themselves, is a deeply revered value in Jain life, and its validity is autonomous and transcends mere self-interest (Wiley, 2006a). Still, it is also true that, according to Jain teachings, when you harm other beings you really harm yourself as well (P. S. Jaini, 1979, p. 167).

But the problem of preventing adhesion is more subtle. The 'stickiness' of the soul to the flowing in of *karma* is not a matter of action as such, though action certainly counts, but of the actor's inner state. What is required is a state of mind in which passions are quelled. One must cultivate a deep equanimity, and even in the midst of life's most severe travails one should remain serene, indifferent to sufferings and pleasures alike, and one should remain controlled and restrained in one's interactions with other beings. The example of the Tīrthaṅkaras, who bore extraordinary hardships and tribulations in complete tranquillity, shows the way. One whose engagement with the world has these qualities will necessarily be 'dry' to *karma*'s influx.

But to liberate the soul from its worldly prison requires more than merely abating karmic influx and adhesion; it is also necessary to rid oneself of karmic accumulations already there. Of course, *karma* does fall away naturally once it has achieved its effect, but in doing so it propels the subject into new actions, and thus the cycle continues. However, it is possible to accelerate the shedding of *karma* by short-circuiting karmic feedback, and the principal means of doing so is ascetic practice.

The point of austerities, when undertaken in the light of Jain teachings and in the spirit of indifference to pain and pleasure, is not self-punishment. Rather, the effects of ascetic practice are twofold. At one level, such practice is both a manifestation of equanimity and a means for its further cultivation, and this lessens one's attractiveness and stickiness to *karma*. But such practice also loosens the hold of karmic deposits directly. The metaphor of fire is often used in this context; austerities are said to 'burn away' the soul's karmic burden.

One might imagine that austerities would be the domain of the mendicant elite, and it is certainly true that a major focus of the lifestyle of Jain mendicants is serious asceticism. Still, Jain life is remarkable for the extent to which lay Jains also engage in ascetic practice. In Jain tradition, ascetic practice tends to be closely linked with food. Food is not only a source of pleasure, itself suspect in the perspective of the religiously serious, but is also the fuel of the body, and the body is the crude outer layer of the soul's material prison. Fasting is only one of many ascetic modes in Jain life, but considering the direct relationship between the conditions of bondage and nourishment (as well as the quest for nourishment which inevitably involves harming some forms of life), the centrality of food cannot surprise. This topic will be discussed further in later chapters.

Liberation's stages

Liberation can be seen as the endpoint of a very long journey. Indeed, it is infinitely long, given the fact that the soul has wandered through an uncreated cosmos for all of beginningless time. And although it appears that the higher animals of five senses possess moral awareness of a sort, and even the ability to engage in ascetic practices, the active quest for liberation requires a human body. In light of the infinities that lie at the foundation of the Jain world-view (Chapter 6), one's possession of a human body must be seen as a very brief sojourn on a road that is long indeed. A human birth is, therefore, a prize beyond price and an opportunity for spiritual progress that should not be wasted.

But simply occupying a human body is not enough, for one must begin a long ascent from the condition of dark delusion in which most souls languish. The path upward is a complex matter involving technicalities that need not waylay us. It can suffice to say that one proceeds upward by suppressing passions and removing *karma* in successive layers, and there is a roadmap for this that describes a series of stages of progress, fourteen in number, known as *guṇasthāna*s (stages of [spiritual] quality) (see the lists in Bothara, 2012; P. S. Jaini, 1979, pp. 272–3; Wiley, 2004, pp. 243–4).

Of these, the first is called *mithyādṛṣṭi* (stage of false views), and this is the condition of all unliberated souls until the climb upward begins. The second and third stages are relatively unimportant intermediate steps, but number four is a major landmark. It is called *samyakdṛṣṭi* (correct viewpoint), which is when one acquires faith in the Tīrthaṅkaras' teachings. This is when the actual journey to liberation begins, and anyone who reaches this stage is bound to obtain liberation sooner or later (even eons later). The fifth stage is attained when one takes the vows required of a Jain layman, and the sixth is reached when one takes the far more rigorous vows of a Jain mendicant. Further progress upward is achieved as one rises to ever higher levels of discipline, restraint and elimination of passions. Slipping backward, even into lower forms of life, is still possible up to and including the twelfth *guṇasthāna*, but all uncertainty about the final destination ends at the next stage.

The thirteenth *guṇasthāna* is a great achievement on the upward climb, for it is at this point that one acquires *kevalajñāna* (omniscience) thus becoming a *kevalin* (omniscient being), which is accomplished by eliminating all remnants of the *ghātiyā karma*s, i.e., those obstructing knowledge, energy and perception. One has now become what is called a *sayoga-kevalin*, a still-embodied and active omniscient being.

By *kevaljñāna* is meant a condition in which knowledge is completely freed from inhibiting *karma*. In this state, knowledge is limitless; it is an unmotivated (i.e., altogether free of the desires and aversions that bring about karmic adhesion) and direct (i.e., without any mental activity) apprehension by the soul of itself and all external substances and objects at all times. Such an omniscient being, however, remains embodied, and embodiment may continue for a very considerable time. The Tīrthaṅkaras remain at this stage during their teaching careers until their final death and liberation.

It will be noted that the concept of omniscience is truly central to the Jain outlook on the world and our creaturely situation. It is not only the goal of Jain religious practice, but also the sole foundation of the authority of the Tīrthaṅkaras' teachings. Why are these teachings valid? It is because they are the teachings of omniscient beings, and for no other reason. It also serves as the master premise of the Jains' epistemological system known as *anekāntavāda* (doctrine of many-pointedness, apparently shared by the Ājīvikas; see Bronkhorst, 2013). In contrast with the *kevalin*'s omniscience, in which things are at once understood from every possible point of view, all normal human understandings are necessarily partial and incomplete. But although the validity of our normal understanding of the world is imperfect, it is enhanced to the degree that it accepts multiple and even apparently contradictory perspectives.

At the instant of the *kevalin*'s death, he or she passes through the stage of *ayogya-kevalin* (inactive *kevalin*). At this point, the body's activity ceases and the very last karmic residues are shed. These are the four *aghātiyā karma*s, those that generate future body, feeling, longevity and environment. The *kevalin*'s soul momentarily expands to fill the entire cosmos and then contracts again; this happens in a flash, and has the effect of spatially offsetting the time remaining on the clock of his *aghātiyā karma*s.

Now the *kevalin* is freed from all sources of bondage and becomes *siddha* (liberated being). The state of liberation is not considered a *guṇasthāna* because it is not a stage on the way to anything. Freed from all former karmic restraints, the soul rises to the zone at the top of the cosmos that is the abode of the *siddha*s. The soul's inherent attributes of bliss, energy and consciousness have now reached their limitless potential in the absence of interfering karmic adhesions. It will now exist for all of infinite time to come in a passionless but blissful state of complete detachment from the world of rebirth below.

The liberated soul retains the shape it had in its final lifetime but is only two-thirds its former size. It does not carry gender, which is shed with the *gotra* karmas that determine such traits. Of course this is not an issue for the Digambaras who maintain, unlike the Śvetāmbaras, that it is not possible to attain liberation in a female body. Because time is without a beginning, the number of souls in the abode of the liberated is infinite, and because the abode of the liberated does have an actual size, as do the *siddha*s themselves, the question arises of how they could all fit in the abode of the liberated. The answer is that they overlap but they nevertheless also retain their individuality.

Given the numerous and extraordinary difficulties of its attainment, liberation is a relatively rare event, even on a cosmic scale. Nevertheless, new souls will continue to enter the liberated state for all of time to come. Because we (in our corner of the world) are in a place and time in the cosmic cycle in which there are no Tīrthaṅkaras currently active (see Chapter 6), nobody will achieve liberation here until the very distant future. For the present, the sixth *guṇasthāna* is the limit. However, even now Tīrthaṅkaras are active in other parts of the world, and so liberation continues to be achieved there. This means that if liberation is currently impossible here, it is indeed possible if one gains rebirth in those other areas.

One might imagine that even though the journey to liberation proceeds at a very slow pace, it would, given infinite time in which to do so, empty the cosmos of unliberated beings. But this would be to ignore the fact that some souls are simply incapable of achieving liberation and the additional fact that the number of souls in the cosmos is infinite. The slow and arduous climb out of bondage will, therefore, never end, just as bondage itself will never cease to be.

Chapter 4

Strivers

Homage to the *arihaṃtas*
Homage to the *siddhas*
Homage to the *ācāryas*
Homage to the *upādhyāyas*
Homage to all the *sādhus*
This five-fold homage destroys all demerits
and of all auspicious things, it is the foremost.

At a very early stage in my work among Jains, I came into contact with the Prakrit formula given in translation above. In fact, I was required to commit it to memory. The reason is quite simple: this is the most important prayer in Jainism. Known as the *namaskāra* (or *navkār*) *mantra,* or *mahāmantra*, it is Jainism's universal expression of reverence and also an encapsulation of one of Jainism's most important principles. A *mantra* is a sacred verbal formula. This *mantra* is a *namaskāra* (salutation) and expression of homage to entities known as the five *parameṣṭhins* (supreme deities), who alone are considered truly worthy of worship in Jain tradition. The *arihaṃtas* (or *arhats*) are those who have achieved omniscience. The *siddhas* are the liberated beings. The *ācāryas* are the leaders of Jain mendicant communities, the *upādhyāyas* are the teachers within these communities, while the *sādhus* are the ordinary mendicants.

Two interrelated points must be made about the *namaskāra mantra*. The first is that these five entities deemed fully worthy of worship by the Jains do not include gods and goddesses. The second is that for all the differences between them, they share one absolutely crucial characteristic: are all mendicants who have renounced the world in favour of the ascetic life. This point is fundamental to understanding Jainism. Jains worship mendicants – the Tīrthaṅkaras most of all, but all mendicants in principle – who alone are regarded as fully worthy of veneration and emulation. And as for gods and goddesses, it is true that they possess extraordinary powers, but in the final analysis they are magnified versions of ourselves, for they, too, are worshippers of the Tīrthaṅkaras.

As we have seen, Jain society is idealised as a fourfold order consisting of monks and nuns and laymen and laywomen. This fourfold order is established and forever re-established by the Tīrthaṅkaras, and, together with Jain teachings, is in fact the *tīrtha* (ford) over the ocean of existence created by them. Indeed, it is their role as 'establishers' that is the defining feature of the Tīrthaṅkaras' teaching mission.

Within this social order, monks and nuns – the mendicant class – are a small minority and are seen as a spiritual elite. Although Jainism gives full sanction to lay life, and while the lay Jains are obviously indispensable to the provisioning and sheltering of mendicants, the fact remains that Jain teachings put the accent on the mendicants' roles as the spiritual leaders of the community as a whole and exemplars of the highest Jain values. As objects of emulation, as teachers, as exhorters and scolders, they are also the principal means by which Jain teachings are imparted to the laity. And in their persons they are also objects of veneration and worship.

The great vows

Foremost among the foundations of the mendicant life are the five *mahāvrata*s (great vows) to which all mendicants commit themselves at the time of their initiation. These vows pledge the mendicant to a manner of living that is the most direct path to liberation (however distant the future in which liberation will actually be achieved). It is also a manner of living that places mendicants in a position of total dependency on the laity. This dependency, which guarantees close and daily interactions between laity and mendicants, is arguably the principal source of the Jain tradition's great strength and ability to preserve and transmit itself successfully, despite the many challenges it has faced over the two and half millennia of its existence.

The *dīkṣā* (initiation of a mendicant) is seen as a rebirth into a new status and social role. Each sectarian group has different initiation formalities, but the essence of initiation remains the same: the renunciation of the initiate's former worldly life, and the adoption of a pattern of life focused on maximum avoidance of harm to living things, severe and sustained austerity, and the cultivation of a permanent state of equanimity. (On initiation, see Vallely, 2002, pp. 77–114; Cort, 1991; Mehta, 2004, pp. 497–540; Shanta, 1997, pp. 457–66; Agrawal, 1972).

Among image-worshipping Śvetāmbara Jains, initiation is a two-step process. Having been feted and feasted for several days prior to the first ceremony, the initiate is dressed in the finery of a groom or bride and taken in procession to the place of the ceremony. There the initiate sheds the finery, puts on the clothing of a mendicant and (in seclusion) has a portion of his or her hair removed. The initiate's hair is supposed to be removed in five handfuls, but in most cases nowadays it is mostly shaved off in advance, leaving a tuft to be pulled out. The removal of hair is part of the initiation process in all Jain traditions, and is understood to be in imitation of Mahāvīra, who did the same when he renounced the world. It can also be seen in more general terms as a particularly vivid way of symbolising the completeness and finality of the initiate's cutting off from his or her former life. The initiate also receives a new name at this stage. Then ensues a probationary period of study and austerity lasting a month or so, during which the initiate will study the scriptures especially relevant to monastic discipline. Then the main initiation occurs, in which the initiate commits himself or herself to the five *mahāvrata*s.

Among Digambaras, initiation involves the movement upward through a system of graded ranks of mendicancy. In the past, and in contrast to the northern Śvetāmbaras, Digambara monks quite often began their mendicant careers late in life, and were frequently recruited from agricultural backgrounds (Carrithers, 1989). The current trend, however, is in the direction of younger and better educated recruits, often from the North.

The Digambara aspirant to full mendicancy (not available to females) is at first a *kṣullaka*. At this level, he wears three items of clothing: a loincloth, a wrap around the waist and a cloth over his shoulders. At the next level, he is an *ailaka*, and wears only a loin cloth. Having passed through these preparatory stages, he can be initiated as a nude *muni*. At this point his hair is removed and he takes the five *mahāvratas*. Having renounced all property, including his loincloth, he now he carries only a water pot and the peacock broom with which he will remove small forms of life from where he sits or lies. A woman of the junior grade is a *kṣullikā*, and a female mendicant highest grade is an *āryikā*. However, because women cannot practise nudity (a cultural fact beyond the reach of challenge), it is impossible for a woman to become a full mendicant in the Digambara tradition; female mendicants, therefore, are seen as spiritually advanced laywomen.

The most important feature of mendicancy among Jains is the declaration of adherence to the five *mahāvratas*. The vows are the same for all sects and subsects and for male and female mendicants. They are negative – i.e., the vow is not to do a given thing – and explicitly spell out both means and modes of application. The aspirant pledges not to do the thing in question through the three means of mind, speech and body. He or she further pledges not to do the thing in three modes – i.e., not do the thing, not to cause it to be done and not to approve of its being done.

The first of the *mahāvratas* is the vow to avoid violence and not to harm any form of life. In many ways this is the most important of the five because it is an expression of a value – *ahiṃsā* – that is arguably the most central of all to Jain life, both lay and monastic, and that motivates and shapes behaviours that are among the best-known features of Jainism. But while lay Jains are also expected to avoid harming forms of life, the prohibition is much wider in scope for mendicants and includes even microscopic one-sensed beings. The aspirant pledges to refrain from causing injury to any living thing – minute or big, mobile or immobile – by means of mind, speech or body, and also pledges neither to do so, nor to cause anyone to do so, nor to approve of anyone doing so. The aspirant also renounces any such injuries inflicted in the past and promises never to do so again. This same basic format is followed in all of the five *mahāvratas*.

The commitment to extend the limits of *ahiṃsā* to all forms of life gives rise to some of the most noteworthy of the many restrictions observed by mendicants. Vegetarianism is, of course, a simple given, but the matter goes well beyond that. For example, mendicants should drink only boiled water, lest they be responsible for the harm done to microscopic beings in the water. Small forms of life swarm in the water, but also the water itself is alive because it contains *āpo-kāyika* (water-embodied life forms). Boiling the water kills all life in the water, including the *āpo-kāyika*. The mendicants are not responsible for these deaths because laity do the boiling.

Mendicants must eat only properly cooked food without living things in it; raw or half-cooked food is forbidden. They must carefully inspect the food they eat to be absolutely sure that there is no possibility of hidden life. Even if it were permitted on other grounds, it would be impossible for mendicants to prepare their own food, for they cannot use fire. This is because fire not only harms life forms in its vicinity, but also is itself alive because it contains *tejo-kāyika* (fire-embodied life forms), and the igniter of a fire would be responsible for their death. For the same reason, mendicants cannot switch electricity on or off, for electricity is understood to consist of *tejo-kāyika*.

The first *mahāvrata* also controls a broad range of behaviour outside the realm of eating and drinking. Mendicants must avoid walking on ground where vegetation grows or where there might be other living things. They cannot dig in the ground because of *pṛthvī-kāyika* (earth-embodied life forms). Śvetāmbara mendicants carry small brooms made of soft wool with which to avoid harming tiny living things on surfaces on which they sit or lie by gently brushing them away. The Digambara version uses peacock feathers (naturally dropped) instead of wool. Mendicants cannot ride in vehicles because of the damage to life they cause. They cannot bathe because of the harm done to water-bodied and water-borne life, but even if they used water devoid of life it would still sink into the crevices in the ground killing life forms swarming there. Bathing also beautifies, which is objectionable on grounds of worldliness. Mendicants must empty urine pots or defecate where there will be no harm to life.

Mendicants must also take special precautions in order to avoid harming airborne forms of life with their hot breath. As is the case with water, living things not only abound in the air, but also the air itself is considered alive because it contains *vāyu-kāyika* (air-embodied life forms). Mendicants belonging to the image-worshipping subgroup of the Śvetāmbara branch carry mouth cloths that they hold in front of their mouths while speaking. These cloths are worn permanently by mendicants belonging to the non-image-worshipping Sthānakavāsī and Terāpanthī communities. Mendicants are forbidden to fan themselves in order to avoid harming airborne life. They are permitted to sing, and do so during certain ceremonies, but they are not permitted to clap or count rhythm on their knees.

Another important consequence of the mendicants' commitment to broad-spectrum *ahiṃsā* is the *cāturmāsa*, the four-month rainy season retreat, which begins in the lunisolar month of Āṣāḍh (June/July) and ends in the lunisolar month of Kārttik (October/November). Jain mendicants must normally stay on the move. Travelling in groups of at least two, they should never spend more than a few days at a given place. During the monsoon season, however, they interrupt their otherwise ceaseless travels and establish semi-permanent residence in a single community, the rationale being that travelling about would endanger the innumerable living things that flourish on the ground during the rains.

The prolonged presence of mendicants in Jain communities results in an intensification of the already close interaction between mendicants and laity. Mendicants give daily sermons and are on hand for regular consultation with lay followers, and the result is a general heightening of religious awareness among laity. Local Jain communities are eager to invite especially famous and charismatic mendicants to spend their rainy season among them, and arrangements are often made years in advance.

One could add much more to this brief account of the direct effects of the first *mahāvrata* on mendicant life, but by now readers will have the general idea. When a commitment to *ahiṃsā* is universalised to cover all living things including the simplest and most microscopic, the result is going to be a way of life beset with behavioural stop signs. Jainism fosters an active awareness of the ubiquity of living things in our environment, and from this perspective to live at all is to create a kind of havoc. The point of the circumspection with which the mendicant life is conducted is to minimise the inevitable damage done.

The second *mahāvrata* is not to lie; the third is not to steal. As Dundas (2002, p. 159) points out, the actual phraseology of the non-stealing vow as given in the *Ācārāṅgasūtra* is 'not to take what has not been given', which is commonly taken to refer to ordinary theft but also includes overly long lingering at a particular place or partaking of alms without the permission of a monastic superior. To this I would add that the 'non-given' stipulation seems to mesh with the concept of *dāna* (merit-generating gift), to which we return momentarily. In this perspective, the third *mahāvrata* further guarantees the complete dependence of the monastic community upon the gifting of Jain laity.

The fourth *mahāvrata* is the renunciation of sexual relations. In part, it arises from the Jain insight (apparently arrived at by means of pure deduction) that semen contains vast numbers of living things for the death of which the ejaculator would be responsible (and indeed the sexual fluids of women also swarm with vulnerable microscopic life). But the vow also extends far beyond chastity to include many restrictions on interaction between mendicants and individuals, mendicants or not, belonging to the opposite sex. Among Śvetāmbaras (not Digambaras), touching and even indirect contact is barred, so that a mendicant should not touch an object that is in contact with a member of the opposite sex. Thus, if a man wishes to give an object such as a book to a nun, he must first place it on a neutral surface, after which she will pick it up. Furthermore, a monk or a nun cannot sit in a place where someone who is a member of the opposite sex has sat until forty-eight minutes have elapsed.

The fifth and final *mahāvrata* is that of non-possession (*aparigraha*), and its importance is second only to the vow of non-harm. This vow fits well with the peripatetic mendicant life, for constant movement ensures that there can be no attachment to a particular place. Jain monks and nuns are permitted to own nothing, although they do carry some personal items. Śvetāmbara mendicants (and Digambara nuns and lower-grade male mendicants) of course wear clothing. Śvetāmbara mendicants also carry certain paraphernalia, such as staffs, their brooms, mouth cloths, alms bowls, and certain personal necessities such as eyeglasses. Digambara monks bear only their brooms and water pots. Mendicants may not handle money under any circumstances. However, some important mendicants in effect control large sums because of their influence over well-off laymen.

Life's roadmap for mendicants

In general terms, the mendicant way of life – as ultimately shaped by the *mahāvrata*s – is highly regulated with the overall aim of minimising the sorts of activities and states of mind that bring about the inflow and adhesion of *karma*. The general idea is that any behaviours – including mental behaviours – not conducive to spiritual growth should be reduced to the fullest extent possible (on these basic rules, see Dundas, 2002, pp. 163–73; P. S. Jaini, 1979, pp. 189–91, 247–8).

The regulation takes the basic form of what are known as *gupti*s (curtailments) and *samiti*s (self-regulations). The *gupti*s are three in number, and are aimed at the restriction of one's speech, bodily activity and mental life. The *samiti*s are five: circumspection in walking; speaking; accepting alms; picking objects up and putting them down; and excretion. Three of the five are directly connected to *ahiṃsā*: one should walk with care lest one harm small creatures underfoot; one should pick up and put down objects with care lest one harm small forms of life in the vicinity; and one should excrete with care (i.e., in an area free from living things and the moisture that gives rise to living things) lest one harm forms of life that might be in the way. Care in speaking, by which is generally meant speaking as little as possible, governs both what is said (i.e., truthfulness) and speaking as an activity that draws *karma*, especially speech that is hateful or boisterous. Care in accepting alms means taking only from those qualified to give to Jain mendicants, and doing so without desire. Mendicants, it is said, should regard food as bitter medicine, useful only to fuel the body at minimal levels along the path leading ultimately to liberation.

Mendicants are also expected to expose themselves to many discomforts and bear them with the kind of equanimity that was discussed in the previous chapter. In general, mendicants are completely at the mercy of the vicissitudes of almsgiving, for the rules of monastic discipline guarantee that they are totally dependent on lay largesse for their every need (although it should be added that this largesse is readily given). Furthermore, because they may not use artificial means of transportation of any kind, mendicants must move on foot wherever they go. They are normally not allowed to use footwear, which means they must sometimes travel barefoot for long distances on scorchingly hot pavement. They must sleep on the ground or hard wooden platforms, must endure hunger and thirst without complaint, and must obey the orders of their monastic superiors without hesitation. Digambara monks must endure cold weather without the benefit of any clothing and are allowed to eat or drink but once a day. The prohibition of drinking more than once a day can be a serious hardship in India's hottest weather. Not only must the mendicant bear these and many other discomforts and inconveniences, but he or she should regard them not as something to be endured but rather as a welcome spiritual opportunity.

Although the hardships undergone by Jain mendicants are severe indeed – a point often emphasised by lay Jains when extolling the virtues of their mendicants to non-Jains – one should bear in mind that the point of the privations is not simply to suffer. Rather, the ultimate purpose of such hardships is the cultivation and maintenance of a deep equanimity in which one is indifferent to pain and pleasure alike.

In addition, mendicants must meet, on a daily basis, certain ritual obligations known as *āvaśyaka*s (essentials). These are also recommended to lay Jains but, in fact, are systematically performed only by the most serious laity. For Śvetāmbaras, these obligations are six in number. The first is *sāmāyikā* (the cultivation of equanimity in meditation). While this is ritualised as a daily obligation, Jain mendicants are understood to be in a permanent condition of *sāmāyika* from initiation to the end of their days. The second obligation is the recitation of a hymn in praise of the twenty-four Tīrthaṅkaras of our region of the world and declining epoch. Cort (2002a, pp. 73–4) points out that this hymn is clearly an expression of what is called *bhakti* (religious devotion), and that it shows that devotional worship, far from being a lay deviation from 'true' Jainism – seen as a purely ascetic tradition as exemplified by mendicants – has been a part of mendicant as well as lay practice from at least the early centuries of the Common Era. The third obligation is *vandana* (ritualized homage to mendicants). The fourth is *pratikramaṇa* (ritualized repentance for the harm one has inevitably done to human and non-human others) while the fifth is standing motionless in the posture of *kāyotsarga* (abandonment of the body). The sixth is *pratyākhyāna* (the formal resolve to renounce certain spiritually insalubrious activities and foods for a specified period of time).

It should not be imagined that Jain mendicants are at all reclusive. They cannot be, precisely because of their dependence on laity for their subsistence; indeed, as has already been noted, they cannot even prepare their own food, which means that Jain mendicants must interact with lay followers on a daily basis. Most mendicants – and especially those who are well known and have large personal followings – are public persons, and to be in their vicinity is to be in the midst of constant coming and going. This should not surprise us because Jain mendicants are not thought of as hermits. The Jain world, as has been shown, is a fourfold social order in which mendicants constitute a separate and elite position, but are not seen as external to what is, in effect, an inclusive socio-religious order.

The mendicant persona

From the lay perspective, the image projected by the mendicant community is an amalgam of two separate but related roles. First and foremost, they are exemplars. Lay Jains are exhorted to wish to lead the mendicant life in some future birth. From the perspective of the path to liberation, the life of the layperson is unsatisfactory, and the very presence of mendicants is a constant reminder of that. But mendicants are also teachers, and the very onerousness and extreme circumspection of the mendicant life imparts special legitimacy to their teaching. In part, the teaching takes the form of public sermons, especially during the rainy season retreat. Many lay Jains also develop intensified relationships with a particular mendicant whom they regard as their personal *guru* (religious preceptor).

Jain mendicants frequently admonish and scold lay Jains for laxity in piety, and I am told that some lay Jains are uneasy about contact with mendicants for fear of awkward questions about their diet. Mendicants are also frequently the instigators of ascetic practices (such as fasting) and other ritual observances on the part of laity. In fact, a layperson must make a formal vow of renunciation before a mendicant (or in front of a Tīrthaṅkara image if no mendicant is available) before undertaking a fast; without such a vow, the exercise will be inefficacious. Also, Mendicants frequently inspire laity (especially well-off businessmen) to support special religious or charitable projects financially, such as temple construction or renovation, pilgrimages, hospitals, animal welfare and the like.

Jain mendicants are also, in their persons, objects of worship. Although the *namaskār mantra*, the all-important formula with which this chapter begins, singles out the omniscient beings and the fully liberated as foremost among the worship-worthy, it also includes living mendicants. Lay Jains worship mendicants in a generic rite of obeisance called *guru vandana* (Cort, 2001, pp. 111–13). In Śvetāmbara tradition, this consists of bowing in coordination with the recitation of a prescribed formula. Mendicants are also worshipped in a more personalised rite known as *guru pūjā* (*ibid.*, p. 114). Central to this rite among Śvetāmbara Jains is *vāskṣep* – a mixture of powdered sandalwood and saffron. The layperson touches (if the same sex) the mendicant's feet and the mendicant sprinkles a small amount of the powder on the head and shoulders of the worshipper. In similar contexts, Hindus use liquids, not powder, but Jains use powder because it can be handled without harming the small forms of life that swarm wherever there is moisture.

Organisation

As described in Chapter 2, the mendicants belonging to the image-worshipping branch of Śvetāmbara Jainism have been organised into *gaccha*s (disciplic lineages) for many centuries, and several of these currently exist. Each *gaccha* is a community of male and female mendicants tracing common ritual and spiritual descent through links between preceptors and their disciples. Currently functioning Śvetāmbara image-worshipping *gaccha*s include the Tapā Gaccha, Kharatara Gaccha, Acala (or Añcala) Gaccha (on which see Balbir, 2003), Tristuti Gaccha, Pārśvacandra Gaccha and Vimala Gaccha. Of these, the Tapā Gaccha, especially strong in Gujarat, is by far the most successful these days. To the best of my knowledge, all currently active image-worshipping Śvetāmbara *gaccha*s trace their disciplic descent directly to Sudharma, who was one of Mahavira's eleven chief disciples, while Sthānakavāsī mendicant groups trace their descent to fifteenth-century Loṅkā Śāh. The Sthānakavāsīs maintain that authentic Jain tradition was lost in the fifth century CE owing to the rise of the image-worshippers, and that they have restored true Jain tradition after a long hiatus (Cort, 1991, p. 656).

On the analogy of descent-group organisation as one sees it in the purely social sphere, one might liken the entire Śvetāmbara image-worshipping mendicant community to a 'clan', with Sudharma as the apical ancestor; the separate *gaccha*s would then be analogous to segmentary lineages into which the clan is subdivided, each with its own common spiritual ancestor. As Cort (1991, pp. 656–7) points out, the formation of a *gaccha* is typically conceptualised in relation to spiritual lapses and reform. A charismatic mendicant arises in response to laxity in monastic discipline; having re-established the proper ascetic rigour, he becomes the spiritual ancestor of a new *gaccha*.

To continue with the analogy of descent, the *gaccha*s are further subdivided into sublineages called *samudāy*s. A *samudāy* consists of those who are, by disciplic succession, the disciplic descendants of a particular mendicant, the apical spiritual ancestor. He is typically a charismatic *ācārya* whose career seems to his followers to have represented a 'new beginning' of some sort (Cort, 1991, p. 661). At the lowest level of segmentation are groups called *parivār*s (families) on the analogy of the family or small lineage segment in a lineage system. As Cort also points out, these mendicant institutions reflect the agnatic values prevailing in most of Indian society at large: monks constitute the core of these entities, and nuns are attached to monks' lineages, just as women become attached to the families and lineages of their husbands by marriage.

The total number of Jain mendicants has been rising in recent years (Flügel, 2006). As for specific figures, the total number of Mandirmārgī mendicants in 1999 was 6,843 (1,489 monks and 5,354 nuns), and of these by far the largest number belonged to the Gujarat-centred Tapā Gaccha, with a total of 6,027 mendicants (1,349 monks and 4,678 nuns) (ibid., p. 321). In 2002, the little-studied and much smaller Sthānakavāsī mendicant community was divided into twenty-six monastic communities, some large, some small. The total number of Sthānakavāsī mendicants was 3,331 (559 monks and 2,772 nuns) (ibid., p. 331). The Śvetāmbara Terāpanthīs are the smallest and most monolithic of all Jain branches; they are a single monastic community under the direction of a single *ācārya*. In 1999, they totalled 688 mendicants (145 monks and 543 nuns) (ibid., p. 334). In 1999, the total number of Digambara mendicants of all ranks, male and female, was 960, with 610 males and 350 females (ibid., p. 362).

In addition to fully initiated mendicants, there still exist today a very few functionaries known as *yati*s among the image-worshipping Śvetāmbara Jains. The *yati*s are a type of mendicant who, unlike *saṃvegī sādhu*s (fully initiated mendicants), own property, maintain residence in one place and are sometimes non-celibate. They represent a surviving remnant of the *caityavāsī*s who were once a powerful force in Śvetāmbara Jainism. Indeed, until the mid-nineteenth century, most Śvetāmbara mendicants were of this type. They were also organised in lineages, and the lineage heads were treated as a sort of religious royalty; they were seated on thrones and were conveyed from place to place on palanquins (Cort, 2001, pp. 43–6). On the Digambara side, the *bhaṭṭāraka*s are somewhat equivalent to the Śvetāmbara *yati*s.

The historically recent, near disappearance of the *yati*s and the corresponding shift of Jain mendicancy to full initiation in the *mahāvrata*s is a deeply important transformation, and this may be seen as a late chapter in the long-running battle between orthoprax reformers and the *chaityavāsī*s.

Jain mendicancy strongly favours patriarchal values. As has been seen, full mendicant status is unavailable to women among Digambaras. Among Śvetāmbaras, monks definitely outrank nuns. In the Tapā Gaccha, nuns do not preach (Cort, 2001, p. 47), but this is not so in the Kharatara Gaccha. Kharatara Gaccha monks are very few by comparison with the Tapā Gaccha, so it seems likely that the teaching role has devolved to Kharatara Gaccha nuns by default.

Nuns greatly outnumber monks, as the figures given above indicate, and this has probably been so for centuries. One explanation for the demographic imbalance lies in the system of marriage prevailing among Jains and the upper castes generally in Indian society. In this system, males are permitted to remarry but females are not, and given the uncertainties of medical care in the past, this meant that young women often found themselves stranded in the condition of widowhood, unable to remarry and (if without children) without a viable role

4.1 Three Kharatara Gaccha nuns.
The middle figure is delivering a discourse.

in family life. Such women in Jain communities often became nuns. Also, abused wives sometimes sought relief in mendicancy, and wives who failed to produce children were sometimes encouraged to become nuns. Difficulty in raising sufficient sums (as dowry) to marry a daughter respectably, or having a daughter who has, for whatever reason, become too old to be married easily, are further scenarios that might compel parents to encourage a daughter to seek initiation. But these circumstances have become less likely to propel women into mendicancy, as attitudes towards marriage and gender relationships continue to change. And as Cort (2001, p. 47) points out, the rising age of marriage and a diminution of the social stigma of widowhood have attenuated the problem of socially redundant widows in recent decades.

All this said, it should also be pointed out that many women enter mendicancy willingly and gladly. In a family, for example, it might so happen that a daughter responds positively to the teachings she hears from mendicants. This might be coupled with intense admiration for the nuns she encounters in her childhood. As she grows older, she tends more and more to seek the company of nuns, and ever so gradually it becomes clear to one and all that mendicancy is where she is heading. It should also be pointed out that becoming a Jain nun is not necessarily a bad option for women – even women who are not escaping a bad domestic situation. For many, it has been an avenue to education, an esteemed position in the community and spiritual fulfilment.

Lay-mendicant transactions

As has been seen, mendicants are entirely dependent on lay followers for the most basic necessities of life, among which the first and foremost is nourishment. Eating is a fraught issue in Jainism. From the perspective of the ascetic values that underlie the mendicant life and that have enormous prestige in Jain life generally, there is a distinct sense in which eating is dangerous. It is both a form of sense enjoyment and the very foundation of the corporeal entrapment of the soul in food-consuming bodies. Moreover, preparing food – even the most vegetarian of foods – inevitably involves violence to forms of life. This being so, refraining from eating has a special spiritual significance in Jainism, which is why fasting is central to the religious praxis of both mendicants and lay Jains (especially laywomen).

Still, that much said, even mendicants must eat in order to live. Indeed, in Śvetāmbara tradition even the Tīrthaṅkaras consume food, although the Digambaras take a different view. And, as seen earlier in this chapter, mendicants are forbidden to prepare food because of the violence cooking involves, so they must obtain their sustenance from the lay community. This they do by going to lay Jains on one or more alms rounds per day. Śvetāmbara mendicants receive their food in alms bowls and then return to the place where they are staying, where they consume the food in privacy. Fully nude Digambara mendicants receive food and water directly in their hands, once a day, and consume it while standing on the spot.

Ideally, mendicants should seek alms only from lay households and individuals of the highest piety and orthopraxy, although this ideal is not always fully realisable, and they should not discriminate between households on the basis of wealth or reputation for good cooking. If no Jain households are available, then a household belonging to vegetarian Hindus can suffice. The food given to mendicants must conform to extra strict rules of purity. Among other things, this means that food for mendicants cannot contain such prohibited (though vegetarian) items as tubers, even if the family otherwise consumes such items, and the food must be prepared under conditions of scrupulous cleanliness and care for the welfare of vulnerable forms of life.

There is one more point of very great importance. The food taken by a mendicant should never have been prepared on his or her behalf – or if it has been (which is sometimes the case), he or she should not know of it. This rule has the obvious consequence that whatever sin was occasioned by the food's acquisition and preparation does not fall on the mendicant who receives it because these activities were not performed at his or her instigation. In this connection, we must remind ourselves that, in Jain tradition, the consequences of action redound to those who do it, cause it to be done or approve of it being done. It is not enough, that is, for the mendicant not to perform a problematical action; he or she must not cause anyone else to do it or even be glad that it was done. Thus, the mendicant should approach the quest for nourishment with total equanimity and feel neither gladness at success nor sadness at failure. Failure, indeed, should be welcomed as an opportunity to fast.

These rules ensure that the moral compromises occasioned by food preparation are those of the preparer. And for the preparer this is not altogether a bad thing. To support the monastic community is an important source of *puṇya* (merit), and *puṇya* translates into success and good fortune, either in one's present birth or later on. It is true that, if one seeks liberation, the 'good karma' or *puṇya* must be shed sooner or later, as all *karma* must be; but in the meantime it is a source of worldly felicity and may even help one along the long road to liberation by putting one in a position to renounce the world fully at some future point. In this sense, when food is presented as alms to mendicants, the negatives of its preparation are offset by the positives of donating to the mendicant community.

In conformity with these ideas, mendicants' alms rounds are supposed to be totally random; that is, the mendicant simply wanders from one worthy household to another in a manner that precludes anyone preparing food especially for mendicants. The normal term for alms rounds is *gocari*, which means roughly 'a cow's grazing'. The basic idea is that the cow moves at random, taking only a little here, a little there, always leaving plenty of grass behind. Not only does this prevent food from being prepared on the mendicant's behalf, but also, because he or she takes so little, nobody in the donating household is deprived of food. In theory, the mendicant consumes such gleanings without enjoyment and in the spirit of complete detachment. Some mendicants mix their food with water into a gruel to be drunk in a gulp. Food should be regarded, one is told, as a kind of 'bitter medicine', an unfortunate necessity to keep the body running while one is on the path to final release.

It is of great importance that readers understand that the alms rounds of Jain mendicants should not be seen as a form of 'begging'. The opposite is true. The mendicant has something of great value to offer the householder whose home he approaches. This is the opportunity to earn merit and spiritual advancement by feeding a mendicant. This brings us to a key principle of exchange in South Asian religious cultures that is highly relevant to the feeding of Jain mendicants. The gift of food to a Jain mendicant belongs to a class of merit-generating gifts known generically as *dāna*. Anything can be presented as *dāna*, but in Jain tradition the presentation of food to mendicants is the epitome of such gifting and central to the relationship between householder and mendicant (for a comparative study of *dāna* in Hindu, Jain, and Theravāda Buddhist traditions, see Heim, 2004).

One could say that, in this regard, there is an interesting tension between divergent interests in the exchange relationship between householder and mendicant. On the one hand, the householder wants to give, and the more he or she gives the better, because giving generates the benefits of merit. On the other hand, the mendicant is disinclined – at least in theory – to take what is offered because taking food compromises his or her ascetic practice. Ideally, the mendicant should just stop eating, which is indeed what some mendicants and laity ultimately do by means of the rite of *santhāra* (death by self-starvation), also known as *sallekhanā*. And because the asceticism of the mendicant receiver is vital for the efficacy of *dāna*, the odd circumstance arises that, in theory at least, the lay donor wishes to give as much as possible to the mendicant least inclined to take.

The complications of *dāna* are illustrated by the nature of the donor-receiver encounter. In Śvetāmbara tradition, the mendicant receiver, approaching the door of the donor, says, '*dharm lābh*' (roughly, 'May you have the benefit of religion'). Whether uttered as an invitation to give food or as a parting blessing, it expresses the idea that the encounter is an opportunity for the donor to achieve some sort of benefit. That benefit can be seen in two frames of reference. In one, *dāna* generates the present or post-mortem fruits of 'merit'. But also, as Laidlaw (1995, pp. 322–3) points out, when one's gifts are accepted by a mendicant, this is a signal of ritual purity and spiritual worthiness, for mendicants are supposed to take only from those whose virtuousness is beyond doubt. In this sense, *dāna* does indeed generate an immediate reward, which is a sense in the here and now of spiritual worth or advancement. Jains, however, maintain that nothing will be gained by gifting *dāna* – or indeed any other ritual action – unless it is done in the right spirit, which is supposed to be one of disinterest in any material reward. In another point of potential tension, the rewards will surely come, but only if you do not seek or even desire them.

Jains stress that mendicants never 'owe' laypersons anything. Carrithers (1989, p. 228) captures the spirit of this relationship well when, writing of Digambara mendicants, he makes the following observation: 'Though material calculation suggests that the *muni* is completely dependent on the laity, the occasion of giving dramatises rather his independence and autonomy.' The merit gained by the lay donor is essentially a 'by-product of the *muni*'s self-purification'.

Venerating mendicants of the past

Ascetic practices have two levels of significance in Indic religious cultures. One level is soteriological; here we encounter austerities as the royal road to escape from the bondage of *saṃsāra*. But austerities are also associated with the acquisition of special powers; here, the basic idea is that ascetic practices unleash inner powers that otherwise lie dormant within the practitioner. From the standpoint of quest for liberation in Jainism, such powers and their acquisition are irrelevant and finally suspect, because they can only distract from the quest for liberation. But the same powers have played an important role in Jainism at the popular level. Many of the *yati*s of times past, for example, had reputations as skilled magicians.

There exists a cult of deceased monks who are worshipped precisely because of the extraordinary magical powers they wielded in life and are believed to exercise still from beyond death's door. This phenomenon has been described only for the Kharatara Gaccha, especially strong in Rajasthan (Babb, 1996, pp. 111–36; Laidlaw, 1995, pp. 74–80; Laidlaw, 1985). And while there appears to be nothing similar within the Tapā Gaccha (although see Dundas, 2007, pp. 53–5 on the apotheosis of Hīravijayasūri), one cannot exclude the possibility of similar cults in other *gaccha*s or other branches of Jainism. In any case, even if it is not pan-Jain, the cult is worth reflecting on for what it teaches about the balance between the quest for liberation and worldly felicity in Jain tradition.

Both Kharatara Gaccha laity and mendicants venerate (albeit in their separate ways) certain distinguished mendicants of times passed. The most important of these deceased mendicants are four in number, and they are known as Dādāgurus (grandfather *guru*s). They are not the only such figures. Other monks of the past (and nuns, too) are commemorated, usually in the form of footprint images, and sometimes actively worshipped, but the Dādāgurus are special. They are *ācārya*s of the past (from the eleventh to the seventeenth centuries) who are singled out from others because of their distinction as defenders and reformers of Jainism, and also and especially as miracle-workers and creators of new Jains.

Most temples associated with the Kharatara Gaccha contain images of these figures. There are also shrines dedicated specifically to them. Such a shrine is called a *dādābāṛī* (garden of the grandfathers). *Dādābāṛī*s are mostly built in the country or outskirts of towns and cities (although many have been absorbed by expanding cities these days) – hence the notion of a bucolic 'garden'. All such establishments include a temple containing images of Tīrthaṅkaras. These images are, in theory, the main objects of worship and serve as a reminder to everyone of what is truly important in Jainism. But the real attraction consists of a shrine containing images of the Dādāgurus, with one of the four as the principal object of worship.

The Dādāgurus are traditionally worshipped in the form of footprint images. Touching the feet of a venerated being is a standard gesture of respect and worship in the Indic world, but footprint images also evoke the image of the last traces of someone who has departed. Anthropomorphic images are more common nowadays, but footprint images are always included with the anthropomorphic ones. The Dādāgurus' images are conventionally housed in structures modelled on the funerary cenotaphs that are a common feature of the Rajasthan countryside, and even when housed in conventional temples they are installed under smaller, cenotaph-like structures. These details point to the fact that the veneration of the Dādāgurus is a mortuary cult, which is arguably true of the veneration of the Tīrthaṅkaras as well. Although the Dādāgurus are closely linked with image-worship and specifically the Kharatara Gaccha, their shrines also attract worshippers from the non-image-worshipping Sthānakavāsī and Terāpanthī sects, and (as I have observed) even some Digambaras also patronise these shrines.

The reason for the popularity of the Dādāgurus is the fusion of two quite different attributes in their personae as objects of worship. On the one hand they are mendicants, and in that sense they are Tīrthaṅkara-like and thus worthy of adoration. But on the other hand, and unlike the Tīrthaṅkaras, they are unliberated; and because of their unliberated status, their relationship with worshippers is very different, as are the meanings assigned to that relationship. From an orthodox standpoint, the Tīrthaṅkara cannot and will not intervene in our worldly affairs, because he is devoid of desires and aversions and disconnected entirely from the world of our struggles and suffering. One could say that he is certainly aware of the entreaties of his devotees, for after all he is an omniscient being, but even if he wanted to (which he does not because he is devoid of desires) he could not respond to those entreaties in any way. But because they are unliberated, the Dādāgurus are still in the world of *saṃsāra* and can act within it. And because they are not yet *vītarāga*, they can desire to do so.

Crucially, they also possess magical powers. In their biographies there are numerous accounts of their magical interventions on behalf of their devotees during their lifetimes, and it is for precisely this sort of intervention that their devotees hope. Their powers stem from the fact that they were, in life, ascetic world renouncers. When seen within the frame of reference of the ultimate goal of liberation, ascetic practice is directly associated with spiritual advancement. But in the less exalted context of popular religiosity, ascetic practice is linked with the acquisition of supernormal powers. In the hagiographies of the Dādāgurus, their magical power is referred to as *yogbal* (yogic power) or *tapobal* (ascetic power), and this is the power that they deploy in the service of the wishes of their worshippers.

As mendicants seeking liberation, the Dādāgurus are truly worthy of worship, which gods and goddesses are not. Because they are not yet liberated, they can act on their worshippers' behalf, and the power they deploy is a product of a specifically Jain ascetic praxis. The cult of the Dādāgurus thus legitimises worship for worldly ends in Jain terms, which accounts for their extraordinary popularity in the communities in which their shrines are found.

Chapter 5

Supporters

I had just had my breakfast on a March morning in 2009 in Jaipur when I got a call from a friend. Had I heard the news? It seems that a well-known figure in Jaipur's gemstone world, a highly esteemed elder of Sthānakavāsī persuasion, had undertaken the vow of death by ritualised self-starvation (usually called *santhāra* by Śvetāmbaras). He was not a friend of mine, but he was an acquaintance and someone I had interviewed in the course of research on Jaipur's gemstone business. I considered him to be a very fine gentleman. It seems that he was not in good health, with a serious and probably fatal intestinal problem of some kind. But there was nothing wrong with his mind, and having sought and received permission from his family, he made his vow with a clear understanding of what he was doing and why. He would leave the world without clinging to it and in a state of self-control and equanimity.

My friend and I arrived at his house in the late morning. A crowd had assembled there and visitors were filing by his open sitting-room window. He was inside, reclining on a simple cot and wearing a mouth cloth, greeting his well-wishers. His family members stressed to us that his mood was quite cheerful. People were there, of course, to pay their respects to a friend and colleague during his last days, but they were also seeking what is called his *darśana*. This term, which may be translated as having an 'auspicious vision' of some exalted being or personage, is most commonly used to describe the encounter devotees have with deities when they visit temples. At a minimum, one would enter the temple to 'take' the deity's *darśana*, i.e., to see and in a sense be seen by the deity.

He lasted eleven days. On the sixth day he took a vow of silence and donned a monk's clothing. Within a day or so after that he entered a coma which lasted until he was gone. His funeral was not the lugubrious affair that one might expect. Rather, the atmosphere was celebratory. The procession to the burning ground included women (not normally a part of funeral processions) and was accompanied by an elephant, several horses, and a truck bearing singers with the inevitable powerful amplifier and speakers. It was a celebration of a life well led and a victorious death.

I have begun this chapter on Jain laity with the incident just related because it encapsulates much that is true about lay life in the Jain world. Jainism certainly permits the pursuit of auspiciousness and worldly successes, despite the importance of ascetic values. This point has not been adequately emphasised in older scholarship, but more recently has been widely recognised, largely as a result of the writings of John Cort (esp. 2001). There is evidence of this in the festive dimension of the funeral and also in the fact that the event presented an opportunity for members of the community not only to reflect on matters of life and death but also to pursue the rewards of meritorious conduct by taking part in the occasion. At the same time, the very fact of a layperson choosing to end his earthly life in this manner shows that ascetic practice is by no means the domain of mendicants alone. It is very much a part of lay Jain life. And as the incident also exemplifies, ascetic and worldly values can be quite closely juxtaposed in Jain ritual practice.

A Jain layman is usually called a *śrāvaka* (the feminine equivalent being *śrāvikā*); the more formal term is *upāsaka*. The term *śrāvaka* means 'listener', which is to say someone who listens to Jain teachings – as gods, humans and animals did in the *samavasaraṇa* of old – and incorporates them in his or her life. By far the majority of Jains of all sects belong to this category.

This chapter explores the religious culture of lay Jains, both as idealised in texts and the sermons of mendicants and as realised in actual life. It begins at the idealised end of the continuum, where a highly elaborated model for Jain lay life is to be found. This model is based on the *śrāvakācāra*s, a genre of mendicant-authored books on ideal lay behaviour that is both ancient and represented by many contemporary works written in modern Indian languages (for a survey of the classical *śrāvakācārās*, see Williams, 1983).

The roadmap of lay life

At the foundation of the idealised version of a layperson's life are two intersecting systems. One is a list of eleven stages of world renunciation through which a layperson should – ideally – rise through life. The other is a list of vows that a layperson should – again, ideally – take. This entire scheme can be seen as an attenuation of mendicant values designed to be consistent with the practical requirements of life in the world.

The eleven stages of world renunciation—known as *pratimā*s—are successively higher levels of restraint tending towards, but not actually reaching, the level of initiated mendicants (for more details, see J. Jain, 1975, pp. 89–90; P. S. Jaini, 1979, pp. 160–85; Wiley, 2004, pp. 151, 245).

The first stage is known as *darśana-pratimā*, the stage of 'correct viewpoint', and it signifies a basic acceptance of the Tīrthaṅkaras' teachings. In Digambara tradition, one will have adopted the most basic norms governing the behaviour of Jains known as the eight *mūlaguṇa*s (basic restraints; P. S. Jaini's rendering) before entering this stage. These govern matters of diet, and point to the fact that dietary rules based on the norm of non-harm are the foundation of the lifestyle of Jains and are also basic to Jain identity. The first of these rules, and the one most basic to Jain daily life, is abstinence from meat. A non-vegetarian is, virtually by definition, a non-Jain. The remaining rules require abstention from alcohol, honey and five types of figs. These other prohibitions are simply elaborations on the basic premise of vegetarianism. Figs, for example, are filled with multiple souls embodied in the seeds, and therefore cannot be consumed without inflicting unacceptable levels of harm to forms of life. The term *mūlaguṇa* has a different meaning for Śvetāmbaras (below), but they have the same dietary restrictions

The next stage involves the system of vows prescribed for laymen that are discussed in the next paragraph. After that, the remaining levels involve successive intensifications of the basic lay vows already taken. Further dietary restraints set the stage for partial and then complete renunciation of sexual relations; one then relinquishes one's profession and family ties, setting the stage for the final *pratimā* in which one is poised at the threshold of initiation as a mendicant.

The second stage of spiritual growth, known as the *vrata pratimā* (*pratimā* of vows) requires the layperson to restrain his or her behaviour by means of a system of vows. These are called *aṇuvrata*s (lesser vows) and are weakened versions of the *mahāvrata*s, the 'great vows') that mendicants must take at their initiations. In general, vows that allow of no exception for mendicants become situational and context-dependent for laity.

The first is the vow of *ahiṃsā* (non-harm). In the case of mendicants, this vow expands to include all forms of life including the most microscopic. However, in recognition of the fact that lay Jains must engage in a vast range of worldly activity that inevitably harms smaller life forms, the lay vow requires lay Jains to refrain from any deliberate violence on forms of life with two senses or more and any unnecessary harm to forms of life lower than that. This vow has obviously had an impact on the occupations into which Jains have migrated. Jains have favoured traditional occupations in which the potential for physically harming forms of life is minimal, and among these running a business stands out as the most favoured of all. Jains do indeed engage in a range of other occupations, and the Digambaras of Maharashtra and Karnataka are apparently mostly farmers, but the Jain merchant is both an Indian stereotype and a reality at the heart of the country's economic life.

In parallel with the *mahāvrata*s, the second lesser vow is that of truthfulness. Whereas a mendicant cannot speak untruthfully on any matter whatsoever, the layperson is given some latitude to speak untruthfully for the sake of minimising harm to others, and is also enjoined to avoid utterances that cause pain or humiliation to others. Given the importance of commerce in Jain life, an important domain of application of this vow is honesty in trade.

The third lesser vow is that of not stealing. The equivalent *mahāvrata* requires mendicants to take nothing that is not given by someone else as a gift (i.e., essentially that which is not given as *dāna*, explained in the previous chapter). While the lay vow also requires the layperson to accept only that which is given, the concept of 'given' is expanded to include such acquisitions as the profits of honestly conducted tradecraft. Also permitted is the taking of such valueless things as dirt or dust if it is done without the intent to steal and if there is no serious loss to the owner. Obviously, there is much potential scope for the application of this vow to the life of the trader.

The fourth lesser vow is that of chastity, but it is, as are the others, an attenuated version of its *mahāvrata* equivalent. One might say that mendicants practise both celibacy (they do not marry) and chastity (they do not engage in sexual intercourse). Jain laymen and laywomen are supposed to be reasonably chaste (for obviously there could be no Jain community if all Jains were perfectly chaste) but not celibate. The layman (the vow assumes a male subject) should avoid sexual activity outside of marriage and in general should practise sexual restraint. For example, one should ideally have sexual intercourse only to produce male issue and should cease sexual activity thereafter. As P. S. Jaini (1979, p. 176) points out, the sexual issue for the Jains is primarily a matter of non-harm. As noted earlier, the sexual fluids of men and women alike are understood to contain vast numbers of microscopic beings that are destroyed by sexual activity.

The fifth and final *aṇuvrata*, the vow of non-possession, is a particularly interesting one in light of the Jain predilection for trade and in light, also, of the ostentation of the lifestyle of many well-off lay Jains (Norman, 1991). At the lay level, the vow is weakened to refer to attachment to possessions rather than possession as such. One is allowed to own property, but within reasonable limits set by oneself. One should avoid the tendency to seek loopholes around these limits (such as shifting property to one's wife's name) and, in general, lead a simple life in which surpluses are donated to religious charities.

The five *aṇuvrata*s are supplemented by three *guṇavrata*s (subsidiary vows). Speaking generally, the purpose of these vows is spatial and temporal restriction of activity and further restrictions of diet. In the first of these vows, one sets limits on the distance one is prepared to travel in given directions. The basic idea is that the greater the physical area and duration of one's activities, the greater the likelihood of bringing harm to living things. In the second, one places further limits on the kinds of things one does, eats or drinks. This vow includes prohibition of cooking or eating after dark (when insects might fall unobserved into the food), which is often taken by Jains themselves to be emblematic of serious orthopraxy. The third is a catch-all, covering a miscellany of misdemeanours.

In addition to the *aṇuvrata*s and *guṇavrata*s are the *śikṣāvrata*s (vows of instruction) – four in number. Their focus is certain ritualized practices, the idea being that one vows to perform a given practice at a specified level of frequency. The first of these is essentially a stricter but temporary version of first of the five *guṇavrata*s; one pledges to confine one's movement to an even smaller physical space. The second, called *sāmāyika*, is a vow to cultivate equanimity in meditation for a period of forty-eight minutes. When a layperson does so, he or she becomes a mendicant for that length of time; mendicants are understood to be permanently in this state. The third is a vow to perform certain fasts, and the fourth is a vow to give alms (specifically *dāna*) to mendicants or spiritually worthy laypersons.

I can certainly understand the exasperation of a patient reader of the above who must now be told that this scheme is mostly pure idealisation and certainly should not be read as an ethnographic description. Few lay Jains actually know and understand this system, and certainly few indeed try to put it into practice. To some small extent it can be seen as a ratification of an already existing reality. Thus, for example, few Jains actually know about the eight *mūlaguṇa*s as such, but what they do know is that these things are wrong, which they have learned in family settings as part of their ordinary socialisation as Jains. Many of the other higher-level prohibitions, such as not eating root vegetables or not eating or cooking after dark, are absorbed (if not acted upon) in the same way. Seriously committed Jains do indeed undertake the vows provided for in the scheme, usually at the instigation of mendicants, but few do so in the context of the whole system, seen as a viable roadmap for living a lay life. It is best to interpret the scheme not as a template but as an expression of certain ideals – obviously mainly emanating from monastic sources – that have broad relevance to lay Jain life in a variety of domains.

Worship

We now turn to patterns of devotional worship among Jains. In so doing, we move from the realm of ideals to that of actual behaviour, for worship is one of the most important realities of normal Jain life. All Jains venerate the Tīrthaṅkaras. It is true that only some Jains worship them in the form of images in temples, because anti-image-worshipping reform movements have enjoyed impressive success among Śvetāmbara Jains. (Such reform has been only marginal among Digambaras.) But even among the Śvetāmbaras – and thus in the Jain world as a whole – image-worshippers are in the majority, and the construction of temples housing Tīrthaṅkara images is a very old Jain practice and one of the most visible manifestations of Jain life throughout the Indian subcontinent.

But herein lies a puzzle. The idea of devotional worship suggests that there is some sort of communication or connection between a worshipper and the object of his or her devotion. However, the most important objects of worship in Jain tradition are the Tīrthaṅkaras, and the Tīrthaṅkaras were not and are not divine beings, at least not in the usual sense of that expression. In life, they were human beings (although of an exceptional sort); after death, they became liberated beings. In liberation, they are *vītarāga* (devoid of all passions) and are entirely inactive within, and indeed totally disconnected from, the world in which we find ourselves. They are self-sufficient and self-contained; they enter into no relationships with worshippers and cannot act in any way on their behalf. What, then, does it mean to worship them? How can they be 'present' in images? More fundamentally, how can worshippers connect with them at all?

Despite these issues, it is quite clear that Jain worshippers do indeed treat the Tīrthaṅkaras as objects of *bhakti* (religious devotion) and worship them in a devotional idiom that freely uses the language of connectedness and entreaty. There is, in fact, a vast body of verse, prayer and song in Jain tradition that simply assumes that the Tīrthaṅkaras are responsive to supplication and 'present' in images, and that portrays worshippers as the objects of the Tīrthaṅkaras' compassion and recipients of their grace (Cort, 2005; Granoff, 1998; Kelting, 2001).

One way to resolve the apparent contradiction would be to assume that devotional worship is a lay contamination from Hinduism of what was originally a purely ascetic path as embodied by mendicants. Indeed, this position has not only been taken by many Western interpreters of Jainism, but is the central claim of the aniconic reformist sects. But, owing largely to the writings of John Cort (2002a, 2006b), we now recognise that devotional worship is not only a key dimension of contemporary Jainism but is also an ancient pattern in which mendicants as well as laity have been deeply involved from the start. Archaeological and literary evidence show that image-worship coupled with devotional sentiment has been part of Jainism from at least the early centuries of the Common Era, and also that images were made and installed at the behest of mendicants. But the matter does not end there, because, despite the importance of devotion in Jain belief and practice, the special status of the Tīrthaṅkaras as liberated beings has also exercised a steady pull on Jain interpretations of the meaning of worship.

We shall return to these issues later, but let us first have a look at Jain image-worship in practice.

Image-worship

The image-worshipping Jains believe that their earthly temples and the practices of worship within them are but imperfect reflections, in our time-bound human world, of eternal archetypes existing in other parts of the cosmos that are exempt from the passage of time and where gods and goddesses have worshipped Tīrthaṅkara images from beginningless time and will continue to do so forever (Cort, 2010b, Chapter 2). These considerations point to an important fact about gods and goddesses among the Jains, and one that must be clearly understood. The Jains credit the gods and goddesses with extraordinary powers and immensely long lifespans, but these deities are not the focus of Jain worship. They are not included, as has been seen, among the five *parameṣṭhin*s, and the latter alone are fully worthy of worship to the Jains. In this respect, the gods and goddesses are our fellow worshippers, for they venerate the Tīrthaṅkaras just as human beings do.

Still, the gods and goddesses do have a special status as worshippers. This is because their rites serve as a charter for human ritual, with humans assuming the personae of deities when they worship images. In particular, men and women take on the roles of Indra and Indrāṇī, the king and queen of the gods. Whenever a Tīrthaṅkara-to-be is born, these two deities, assisted by many others, give the infant his first post-partum bath on the summit of Mount Meru. The bathing of a Tīrthaṅkara image – basic to Jain ritual culture – is seen as an earthly re-enactment of this event, and Jain worshippers sometimes wear crowns as a way of signifying their identity with the divine kings and queens.

As Cort (2010b, pp. 67–98, 109–11) points out, the Jains do not place their divine archetypes of worship at the beginning of time (as

5.1 A crowned worshipper.

do other traditions) because time has no beginning for the Jains. The placement of divine charter-rites must, therefore, be remote in space rather than time. Thus, we find the gods worshipping imperishable Tīrthaṅkara images positioned in various regions of the cosmos inaccessible to us, but especially at Mount Meru, the centre and axis of the world.

At the same time, there is an alternative ritual archetype applying especially to temple worship. When a Jain layperson goes to a temple, this is seen as analogous to a visit to the *samavasaraṇa*, the structure and assembly within which Tīrthaṅkaras address gods, humans and some animals. In a sense, this represents a beginning in time, but in relative rather than absolute time. It is a moment – one moment among an infinity of such moments – of 'establishment', for it is from the *samavasaraṇa* that the Tīrthaṅkaras create the 'fords' that make liberation possible.

Images of the Tīrthaṅkaras are remarkably alike. They depict stylised human figures in a meditational pose, standing or seated. For the most part, the images of different Tīrthaṅkaras are indistinguishable except for a small symbol at the base of the image. Mahāvīra's symbol, for example, is the lion. However, Pārśva, the twenty-third Tīrthaṅkara of our epoch and region of the world, is mythically associated with the cobra, and his image is typically surmounted by spreading cobra hoods.

A highly conspicuous and important difference between Śvetāmbara and Digambara iconographic tradition is that Śvetāmbaras equip their images with glass eyes and other adornments, whereas the Digambaras do not. The refusal of the Digambaras to decorate their images makes sense in light of the centrality of nudity in their conception of the Tīrthaṅkara, for they maintain that one cannot truly be a monk without relinquishing clothing, to say nothing of being a Tīrthaṅkara. But there is also a deeper difference (Cort, 2010a). The Śvetāmbaras maintain that eyes help the worshipper to visualise himself or herself in the Tīrthaṅkara's presence as an 'as if' experience. The Digambaras say that the true function of a Tīrthaṅkara image is to inspire the worshipper to turn inward and away from the external world, and that eyes and other decorations interfere with that by drawing consciousness outward.

5.2 A fully decorated image of Parśva in a Śvetāmbara temple.

A Jain temple typically has a principal Tīrthaṅkara image plus side shrines housing images of Tīrthaṅkaras other than the main one. Some of the Tīrthaṅkara images on the altar are small and portable to allow their use in special ceremonies. A temple is typically referred to by its main image: such and such a Mahāvīra temple, or Pārśva temple, and so on. Temples exist for all twenty-four of the Tīrthaṅkaras of our declining epoch and place in the world, but of these the most popular are probably Pārśva, Mahāvīra, Śānti (number sixteen of the twenty-four on whom see Cort, 2001, pp. 197–9) and Ṛṣabha (the first).

Certain subsidiary figures and guardian deities are also commonly represented by images in temples. Subsidiary shrines for the Dādāgurus are usually present in temples associated with the Kharatara Gaccha. Guardian deities of some sort are found in all temples. Their job is to protect the temple premises and the Tīrthaṅkara images therein. Some are linked with particular Tīrthaṅkaras; an example is Dharṇendra and Padmāvatī, a pair of cobras – male and female – whom Pārśva rescued from a fire and who then became his guardians. Some of these godlings – the latter pair among them – have acquired reputations for helping worshippers in worldly matters, but one's dealings with them should be seen as entirely subsidiary to the worship of the Tīrthaṅkaras.

We now turn to some of the specifics of Jain ritual culture. For the most part what follows is drawn from Śvetāmbara image-worshipping tradition, which is the ritual tradition I know best.

Visiting temples

A Jain temple is a place of ritual isolation from the hubbub and contaminations of life outside. Prior to its installation and consecration in a temple, a Tīrthaṅkara image is a mere inert statue, without any special qualities other than its shape. After the *pratiṣṭhā* (consecration ceremony) it becomes a sacred thing and, accordingly, it must be protected from impurities.

When one enters a temple, one should be in a condition of ritual purity, although the level of required purity depends on the nature of one's intended encounter with the image or images. People in a seriously polluted condition (such as menstruating women or those mourning a recent death in the family) should forgo temple visits altogether. Those who enter should leave footwear behind and rinse off their feet and rinse out their mouths before doing so. Ideally, a temple visitor should avoid wearing or carrying items made of leather (such as belts or wallets) in a temple, although this rule seems to be more strictly enforced by the Digambaras than the Śvetāmbaras.

More exacting rules govern those who enter the inner shrine and actually come in contact with the Tīrthaṅkara images. They must be freshly bathed and wearing special clothing reserved for worship. Men's clothing should be unstitched; for women, some stitching is allowed. While in contact with the Tīrthaṅkara images, worshippers should cover their mouths with a cloth to prevent impurities carried by their breath from polluting the images.

With the exception of Digambaras of South India (P. S. Jaini, 1979, p. 195), the role of priests who mediate between worshipper and objects of worship is absent in Jain ritual culture. In the North, Śvetāmbara and Digambara temples alike are served by temple employees who keep the temple clean, prepare materials used in worship and assist worshippers in other ways, but they are not seen as ritual mediators in any sense; rather, they are mere temple servants and come from various social backgrounds.

The frequency of temple-going varies greatly among lay Jains, but observant Jains normally visit temples as part of their morning routine. Morning, indeed, is when the temples are truly alive. Unless there is some special occasion, the only significant event in the evening is a typically sparsely attended *āratī* (lamp offering) ceremony.

A temple visit presents an array of institutionalised but flexible modes of encounter with the Tīrthaṅkara, as represented by his image or images. At a minimum, one might merely stop by briefly to take the image's *darśana*. The *darśana*-taker might simply greet the Tīrthaṅkara image with folded hands and meditate briefly on the Tīrthaṅkara's qualities. Or, after paying homage to the image, he or she might make three clockwise circuits around the image (through a special hallway behind the altar found in most temples). After taking *darśana*, a worshipper might perform the full rite known as *caitya vandana* consisting of oral recitations of praise verses coordinated with a series of physical obeisances.

Some worshippers might perform *snātra pūjā* (a bathing rite) focused on bathing a portable Tīrthaṅkara image brought out from the altar into the temple's main hall. The worshippers take the part of Indra and Indrāṇī giving the newborn Tīrthaṅkara-to-be his first bath on Mount Meru. The image is often placed on a three-tiered stand (normally representing the *samavasaraṇa*, but here standing in for Mount Meru) and the devotees perform the prescribed acts in coordination with the singing of a text that tells the story of the Tīrthaṅkara's birth and first bath.

Very commonly, however, Śvetāmbara worshippers perform a rite called *aṣṭaprakārī pūjā* (the eightfold worship), or some simplified or

5.3 Bathing a Tīrthaṅkara image.

idiosyncratic version of it. This rite is truly fundamental because many other more complex rituals, including a variety of congregational rites known as *mahāpūjā*s, are essentially elaborations on its basic framework. The *aṣṭaprakārī pūjā* is worth discussing in some detail not only because of its ubiquity but also because of what it tells about the relationship between ritual acts and Jain soteriological values.

The eightfold worship

Here I describe the *aṣṭaprakārī pūjā* in the form prescribed in a laymen's manual authored by a Śvetāmbara monk named Muktiprabhvijay (n.d., pp. 48–61). (More complete descriptions and analysis of the rite can be found in Babb, 1996, pp. 84–101; Cort, 2001, pp. 71–80.) The view of ritual expressed in this book is that of a mendicant perspective, but it can also be seen as expressing in systematic form an understanding of the rite shared widely by lay as well as mendicant Jains. Muktiprabhvijay describes the *aṣṭaprakārī pūjā* as a series of eight distinct ritual actions and provides verses to be recited or meditated upon while performing each of them. While not normative, these verses provide a context in Jain teachings for each of the separate acts.

Generally, the *aṣṭaprakārī pūjā* is performed by individuals in the morning in a temple. The ritual that a given individual does might or might not be the entire rite as described here, for there is ample room for individual variations in Jain ritual culture. Some enact the rite in silence, perhaps inwardly reciting verses or prayers. Others recite or sing verses such as the ones supplied by Muktiprabhvijay in his book.

The first of the eight segments of the *aṣṭaprakārī pūjā* is worship with water. The worshipper cleans the image of the remains of previous rites and then pours a mixture of water and other substances over the image followed by pure water. This is the Tīrthaṅkara's post-partum bath, as performed by the gods, and the verse accompanying the act in the laymen's manual expresses the worshipper's desire that the dirt of his or her *karma*s be washed away. The leftover water is considered to participate in the image's sacredness and the rite's performers often anoint their foreheads and eyes with it.

Part two of the *aṣṭaprakārī pūjā* is worship with sandalwood paste. The performer applies a paste of sandalwood to nine places on the image's body, and the verse for this act expresses the desire for the worshipper's anger, lust, etc., to be cooled as sandalwood is cool (sandalwood is considered 'cooling' in South Asian cultures).

Part three is the decoration of the image with flowers. The accompanying verse expresses the hope that, by paying homage with flowers, the worshipper's life will become fragrant with the five-coloured flowers of the three jewels of Jain tradition (right faith, right knowledge, right conduct) plus *tapas* (austerity) and *vīrya* (strength). Using flowers in worship is a controversial act in Jainism because of the violence involved in picking them, and some Jain traditions permit only artificial garlands.

The first three parts of the *aṣṭaprakārī pūjā* as described above involve coming into direct contact with the image. As a group they are called *aṅg pūjā* (limb worship) because of the focus on the image's body and its parts. The remaining five parts of the *aṣṭaprakārī pūjā* are known as *agra pūjā* (worship before [the image]). This part of the ceremony occurs at a distance from the image in the temple hall and does not require the worshipper to come into direct contact with the image, so the need for purification of the performer is less strict, and some temple visitors perform only these five segments.

Having left the inner shrine, the devotee performs the fourth act of the *aṣṭaprakārī pūjā* (and the first of the five 'before' the image), which is worship with incense. While standing, the performer circles burning incense before the image. In Muktiprabhvijay's (n.d.) manual, the verse associated with veneration with incense expresses the performer's wish that doing the rite will purify his or her inner life in the same manner that the fragrance of the incense drives inauspicious influences away.

Next, and fifth, is lamp worship, which – while still standing – the performer does by circling a lighted lamp before the image. The associated verse expresses the performer's hope that his or her ignorance will be dispelled by the lamp of knowledge.

The remaining three parts of the *aṣṭaprakārī pūjā* (six, seven and eight) are arguably the most significant of all from the standpoint of the Jain world-view, for in this sequence of ritual acts some of the most important Jain concepts are symbolised. Each of the three rites consists of an offering made on an elevated surface of some kind. This can be a low table positioned in front of the worshipper while seated on the temple floor, but offerings can also be made on the top of the donation box. This is typically a large box situated in front of the shrine in which the temple's main image is housed, and is slotted on top for donations. The performance of these three rites is a highly conspicuous activity in Jain temples, and something that non-Jain visitors are likely to notice first when they visit Jain temples.

The first, and thus the sixth part of the *aṣṭaprakārī pūjā*, is worship with unbroken rice grains. Using grains of rice, the performer forms a diagram on the surface on which the offerings are being made. At the bottom of the diagram is a *svastika* (an ancient symbol of auspiciousness in India that, unfortunately, most Westerners associate with the Nazi regime in Germany). Above the *svastika* are three small heaps of rice, and above them are an upward-facing crescent with another small pile of rice positioned between the crescent's two arms.

Muktiprabhvijay's (n.d.) laymen's manual provides a highly instructive interpretation of these symbols. The four arms of the *svastika* represent the four great destiny classes of unliberated beings: humans; deities; hell-dwellers; and animals and plants (Chapter 6). The three small heaps of rice stand for the three jewels of Jain tradition, and the crescent and dot represent souls of the liberated in their abode at the top of the cosmos. This figure, taken as a whole, is therefore a representation of the Jains' basic understanding of the nature of the cosmos and our position in it. The *svastika* represents *saṃsāra*, the round of birth and death. The three heaps of rice above allude to the means – and the only means – of escaping *saṃsāra*. The crescent and dot point to the final goal, which is liberation itself.

The diagram now drawn, the next part of this sequence (and the seventh part of the *aṣṭaprakārī pūjā*) is worship with food. Usually sweets of some sort (often rock sugar) are offered by placing them directly atop the vertex of the *svastika*. The symbolism of food offerings is rich indeed. Food is nourishment for the body. As has been seen, this is a property of food that gives it a dubious reputation in Jain teachings, for the body is the prison of the soul. Thus, placing a food offering on the centre of the *svastika* – which represents the arena of bondage and is precisely where food belongs – condenses much of the Jain view of our worldly situation into a single symbolic act.

5.4 *Aṣṭaprakārī pūjā* remnants. Note the diagrams executed in grains of rice.

Connected with this is another crucial point. Although food is offered before the Tīrthaṅkara, there can be no question of the Tīrthaṅkara's partaking of the offering, at least not from an orthodox standpoint, because he is a liberated being. As such, he has left all appetite behind. This being so, the food offering is understood to symbolise the renunciation of food in imitation of the Tīrthaṅkara. Here is the verse supplied by Muktiprabhvijay's (n.d., p. 60) manual in association with the food offering: 'Oh Lord! I have eaten and am weary of eating, and you are non-eating [*aṇāhārī*] and obtain the highest happiness. You are non-eating and I am one who eats [*āhārī*]. By worshipping you with this *naivedya* [a general term for food offered to a deity], I want to obtain the condition of non-eating.' By putting the food on the *svastika*, which represents the world of birth and death, one is putting the food in *saṃsāra*, which is exactly where it belongs and where it must stay (Cort, 2001, p. 78; Humphrey and Laidlaw, 1994, p. 128).

Then, having symbolically shed the burden of alimentation, the worshipper moves on to the climax of the whole ceremony – worship with fruit, which is rite number three of the series and the eighth and final part of the *aṣṭaprakārī pūjā*. The worshipper places fruit of some kind on the crescent at the top of the diagram, which represents the idea that liberation is the desired 'fruit' of the rite, and the verse in Muktiprabhvijay's (n.d.) manual expresses the hope that the offerer will obtain liberation.

The Tīrthaṅkara in ritual culture

As described earlier in this chapter, orthodoxy forecloses the notion that the Tīrthaṅkara is 'present' in images or in any sense at all in the worshipper's world, and yet devotional worship, conducted in the idiom of connectedness, is an important dimension of Jain ritual culture. Most ordinary Jains do not seem to find this a problem. Some do not actually understand that they are not supposed to believe that the Tīrthaṅkara is present. For others, disbelief in the Tīrthaṅkara's presence may be cognitively possible but emotionally impossible – what the mind knows the heart rejects. Still others understand and accept the Tīrthaṅkara's absence; for them it is an 'as if' experience. In explanation, they sometimes compare the Tīrthaṅkara image to photographs that one might have of one's deceased parents. We keep them to remind us of our parents' good qualities, but the photos do not bring them back to life.

That said, from the orthodox standpoint – always exerting a kind of tidal tug on Jain ritual culture – worship is, whatever the heart may say, not really a *relationship* between two entities – worshipper and worshipped – at all. Rather, it is *reflexive*, which is to say that its efficacy consists in how the worshipper modifies himself or herself through the act of veneration. Thus, in the *aṣṭaprakārī pūjā* (as subjected to a learned mendicant's – Muktiprabhvijay's (n.d.) – understanding), the act of worshipping the ultimate ascetic (the Tīrthaṅkara) becomes itself an ascetic act. In making a food offering, the worshipper is not 'giving food to' anyone or anything; instead, the he or she is symbolically 'giving food up'.

But even so, the importance of reflexivity as a theme in the ritual culture of Jainism (again, from the orthodox perspective) does not in itself preclude a strong devotional dimension. This is illustrated by the following passages taken from a lay manual authored by a Kharatara Gaccha nun named Hemprabhāśrī (1977, pp. 25, 27). The performance of *pūjā* (worship), she writes, results in 'purity of soul'. A few pages later, she elaborates: 'Just as the *darśana* [auspicious vision] of the Supreme Soul [the Tīrthaṅkara] makes the mind pure and brings about the removal of *karma*, so the worship of the Lord encourages the arising of *bhāva* [spiritually beneficial feelings], and the spark of these feelings will burn *karma*s and reduce them to ashes. Worship is performed in order to destroy sensual vices and eradicate *karma*. Just as austerity and self-denial eradicate *karma*, worshipping the Lord with devotion also destroys *karma* and provides many worldly benefits besides. Auspicious feelings will result in the adhesion of *puṇya* (merit), and from *puṇya* will automatically come material happiness'.

From Hemprabhāśrī's standpoint, worship is actually a substitute form of ascetic practice, and as such it directly eradicates *karma*. In this, she echoes Muktiprabhvijay (n.d.). But at the same time, devotional emotion has a key role in Hemprabhāśrī'sscenario. Worship with devotion, she also tells us, gives rise to spiritually beneficial *feelings* in the worshipper – by implication, feelings that arise from, and express, one's admiration of the good qualities the Tīrthaṇkara represents. These feelings, too, will burn away *karma*s. And more, worshipping him with devotion can even generate worldly benefits by means of that positive form of *karma* known as *puṇya*.

But we must carefully note what Hemprabhāśrī does not say. Whatever laypeople might know or not know, believe or not believe, nowhere does she say that any benefit of the rite – spiritual or worldly – will be bestowed by the Tīrthaṅkara. In a sense, devotion itself is (admittedly a mendicant's point of view) reflexive; the beneficial feelings arise not from some substantial connection with the object of worship, but from the worshipper's appreciation and love for the qualities his image represents.

The worldly efficacy of ritual

But there are also other angles on the issue of how worship can produce good worldly results. It is often said that the veneration of the Tīrthaṅkaras pleases the gods, especially the guardian deities associated with certain specific Tīrthaṅkaras. Being entirely devoid of desire and aversion, the Tīrthaṅkara is neither pleased nor displeased by the fact that he is being venerated, but his guardian deity or deities may be gratified by the spectacle and reward the worshipper as a result. In this connection, it should be noted that the Jains' concept of three modes of wrong conduct works in reverse. Just as one vows not to do a wrong thing, to cause it to be done or to approve of its being done – for all three are demeritorious – so too to do something admirable, to cause it to be done or to approve of its being done are all meritorious. Thus, even the witnessing bystanders benefit when acts of worship take place, and this includes the deities, especially those who are most pleased to see their particular Tīrthaṅkara being venerated. Many believe that veneration at temples devoted to Ṛṣabha and Pārśva create particularly powerful effects because of the power of two goddesses, Cakreśvarī and Padmāvatī, who are linked with these Tīrthaṅkaras respectively.

In addition, some *kṣetrapāl*s (guardian deities of particular temples') acquire fame as helpful to worshippers in worldly difficulties and have become independent deities in their own right. One such, with a large following in Gujarat, is Ghaṇṭākarṇ Mahāvīr. He is the guardian deity of a Padmaprabha temple (the nineteenth Tīrthaṅkara of the twenty-four) at Mahudi in Gujarat, and such is his fame that this temple is one of the wealthiest Jain temples in India (Cort, 2001, p. 91). He somewhat resembles the Hindu deity Hanumān, who plays a similar role as 'helpful deity' for Hindus, and his image is found in Jain temples and household shrines throughout the Śvetāmbara world. Another very famous deity of the same sort is the guardian deity of a Pārśva temple located in the Rajasthani town of Nakoda. Known as Nākoḍā Bhairav, he has a large following in Rajasthan, and his images are installed in many other Śvetāmbara temples and shrines. Some businessmen have made him a partner in their businesses and have pledged a percentage of their profits to him.

On a point that has already been introduced in the previous chapter, in areas falling under the influence of the Kharatara Gaccha, Śvetāmbara Jains venerate the Dādāgurus with the idea that they can assist their worshippers in worldly matters. No longer of this world but not yet liberated, they possess ascetic power that they are able and willing to deploy for their followers.

These examples of unliberated and helpful deities suggest that a strong commitment to liberation achieved through world renunciation leaves a gap in the religious lives of laypeople that needs to be filled somehow. The trick is to fill it in such a way as not to challenge the transcendent authority of ascetic values. It is probable that without recourse to such deities as Ghaṇṭākarṇ Mahāvīr, Nākoḍā Bhairav or their equivalents, Jains would turn to Hindu deities to satisfy their desire for supernatural solutions to their problems, which indeed some Jains do in any case. The cults of Jain goddesses and powerful guardian deities legitimise this impulse in Jain terms; they provide a Jain context for a type of religiosity that in other respects fits poorly with Jainism's loftier goals. It is worth noting that these helpful deities also attract numerous worshippers from the non-image-worshipping sects.

Ritual and social honour

While the simple *aṣṭaprakārī pūjā* is, for many, a quiet, private affair, quite a number of ritual occasions are public events and congregational in nature. This is true in both Śvetāmbara and Digambara traditions. Examples include various calendrical rites, the consecration of temple images, the initiation of mendicants and many other ceremonial occasions. In such situations, the question arises of who among the laity present will represent the assembled worshippers in performing certain key ritual actions. This question is usually answered by means of potlatch-like auctions (of which the best available study is Kelting, 2009). These auctions are intended to solicit donations for the support of ceremonies and temples, and can also be seen as a device for the conversion of material wealth into religious merit. At the same time, they are an arena for the seeking and validating of social honour on the part of the community's most wealthy and successful individuals.

With someone serving as auctioneer, bids are solicited, and the bidding can be highly competitive in important ceremonies. If the auctioneer feels that the bidding is too slow or too low, he will scold and exhort the bidders to get moving and do the right thing. The bidding process can be quite lengthy, and is watched with keen interest by attendees. This is because its ups and downs can be an excellent source

5.5 Auction at a ceremony.

of social information about the bidders and their families. The well-off few will naturally be under pressure to offer high bids because failure to do so might seem as stinginess, or worse as a sign of declining fortunes, but for a family of lesser status to bid too aggressively can seem presumptuous.

Such donations are examples of the transmutation of economic means into the 'symbolic' capital of high social standing (Bourdieu, 1977, pp. 171–83). And more, because these ceremonial donations revalue business success in the coin of demonstrated piety, an auction winner in a major ceremony projects an image that blends worldly success and good character. This, in a feedback loop, enhances his reputation for trustworthiness in the marketplace, and thus contributes to his business success.

These competitions belong to a religious division of labour based on gender (Reynell, 1987; 1991). Women are the primary caretakers of a family's day-to-day piety, especially in the arena of ascetic practices, for they are the ones who typically do the fasting. They usually come into conspicuous public view in the role of successful marathon fasters who are feted publicly. Only rarely have I seen women take part in ritual auctions. As Balbir (1994, p. 127) points out, women are also arguably the principal 'donors' in Jain society, for it is they who support the mendicant community with food. By contrast, given their role as breadwinners – and given, also, the tight connection between wealth and honour in trading communities – the ceremonial role played by men emphasises their wealth and the material support that they provide for the community's ritual and spiritual endeavours.

Calendrical rites

The observance of periodic celebrations and rituals is an important part of Jain religious life. However, the Jain character of the ceremonial calendar is complicated somewhat by the fact that Jains and Hindus share some festivals (most notably Dīvālī), albeit with different interpretations of the significance of the occasion, and many Jains are also simply drawn into the non-Jain celebrations of their Hindu neighbours. The situation is further complicated by the fact that, although the Śvetāmbaras and Digambaras have generally similar ceremonial calendars, the dating of their *parva*s (holy days) is not quite the same.

Without a doubt, the most important of each year's ritual events for Jains is focused on mendicants. This is the annual rainy season retreat of four months' duration, which begins, as calculated in the lunisolar calendar, in June/July and ends in October/November. This is the period when mendicants must cease wandering and remain in one place so as to avoid harming the many forms of life that swarm on the ground during the annual monsoon rains. The extended stay of mendicants in a community inevitably results in a heightening of religious sentiments and behaviour, and even relatively non-observant Jains tend to cultivate an intensified Jain-mindedness during this period. This is a direct result of the fact that mendicants are present giving daily sermons and exhorting laity to engage in ascetic practices such as fasting and to refrain from such forbidden activities as eating after dark.

Of the events taking place during the rainy season retreat, the most important by far – and indeed the most important event in the Jain sacred year – is what the Śvetāmbaras call Paryuṣaṇa (Abiding). The Digambaras have a similar celebration called Daśalakṣaṇaparvan (Holy Observance of the Ten Virtues). The theme of Paryuṣaṇa is austerity and repentance. Mendicants give daily sermons for the entire period. During this time, laity fast and observe *pratikramaṇa* and *poṣadha*. *Pratikramaṇa*, briefly mentioned in the previous chapter, is a rite of confession and atonement designed to ameliorate karmic accumulations during a preceding period. It is standard among mendicants and carried out by laity on holy occasions. Many Jains perform it on a daily basis during Paryuṣaṇa, and virtually everyone performs it on the final day. The term *poṣadha* refers to a vow in which a layperson temporarily lives the life of a mendicant, and some of the seriously orthoprax take this vow for the entire eight-day period of Paryuṣaṇa.

A major feature of Paryuṣaṇa is recitation of the *Kalpasūtra* and commentary by mendicants (Cort, 2001, pp. 151–9). These sometimes sparsely attended recitations begin on the third or fourth day and take place in a sermon hall in the presence of a lay audience. Readers will recall that this text is one of the core texts of Mandirmārgī Śvetāmbara tradition, and that its main feature is a narrative of Mahāvīra's life.

When, on afternoon the fifth day, the recitation of the Prakrit text comes to the account of Mahāvīra's conception and birth and the fourteen dreamlike visions seen by his mother, there ensues one of the most important events of the sacred year. At this point, the hall has typically filled with laity, male and female, young and old, dressed in expensive finery. Replicas of each of the fourteen dreamlike visions are ceremonially taken, one by one, to the front of the hall, an act that represents the appearance of the visions to the mother. The privilege of presenting each vision is auctioned, and the importance of the occasion and the wealth of many of the attendees guarantee that large sums will be bid. At the end, the congregation worships a model cradle, standing for the infant Tīrthaṅkara-to-be. The atmosphere of gaiety and abundance on this occasion stands in radical contrast to the deeply solemn character of Paryuṣaṇa as a whole.

At the conclusion of Paryuṣaṇa is the performance by laity of the annual *pratikramaṇa* known as *saṃvatsarī pratikramaṇa*. Afterwards, the participants ask forgiveness of each other for any harm they might have caused during the previous year. The verbal formula for this is *micchāmi dukkaḍam* (may my improper acts be without consequence), and on this occasion Jains also send cards and letters to each other requesting forgiveness for any harm done during the forgoing year. Those who have fasted during Paryuṣaṇa break their fast on the following day, and later there is a feast for the entire local Jain community.

While the rainy season retreat is still in process comes the festival of Dīvālī (at the new moon of the lunisolar month of Kārttik, October/November). This festival (actually a cluster of various rites) is arguably the most important of the year for Hindus. Its main focus is the worship of Lakṣmī, the Goddess of Wealth, and because of this, and also because it is the beginning of the new business year, it is of great importance to most Jains as well. Dīvālī is a contraction of Dīpāvalī (Row of Lights), and the festival is so-called because of the custom of placing clay lamps in and around houses and businesses in the evening, the basic idea being (according to one interpretation) to guide Lakṣmī into the home. Most Jains worship Lakṣmī in the evening of Dīvālī along with their new account books. Although Jains participate in these events, they also have their own interpretation of the festival. They consider the midnight of Dīvālī to be the moment when Mahāvīra achieved liberation and believe that Indrabhūti Gautama, his chief disciple, attained enlightenment in the wee hours of the following morning. Although mendicants address these points in sermons attended by the religiously serious, a festive, worldly-felicity state of mind dominates the Dīvālī season for most Jains as well as Hindus.

Mahāvīra's birthday – Mahāvīra Jayantī – is celebrated during the waxing fortnight of the lunisolar month of Caitra (March/April). This is one of the few festivals commonly observed jointly by Śvetāmbaras and Digambaras. It is a public holiday in India, and is often celebrated by processions through urban areas, with floats depicting events from Jain scripture and promoting non-harm and vegetarianism.

In the waxing fortnight of the lunisolar month of Vaiśākh (April/May) comes Akṣaya Tṛtīya (Immortal Third). This commemorates the first feeding of Ṛṣabha, the first Tīrthaṅkara of our epoch and corner of the terrestrial world. As the story goes, after Ṛṣabha's renunciation of the world he remained without food for thirteen months because the people of that era had never encountered mendicancy before and had no idea of the proper way to feed a monk. At last, King Śreyāṃsa of Hastināpur, his grandson, recalled the proper method of feeding a monk from a previous birth and offered Ṛṣabha some sugar cane juice. This was the first act of feeding a mendicant of our declining epoch and place, and given the absolute centrality of the mendicant–lay tie in Jainism, it was an important event indeed. Temple images are bathed with sugar cane juice, and ascetics and laity who have completed a special year-long sequence of fasting are given sugar cane juice by devotees and family members.

There are many other observances and rites of importance to Jains embedded in weekly, fortnightly, monthly, four-monthly and annual cycles of time. In addition, periodic initiations, image consecrations and annual commemorations of the image installations in temples are also a major part of the ritual life of Jains.

Giving up

For lay Jains of all traditions, ascetic practice is one of the pillars of religious life, which is to be expected in a religious tradition to which ascetic values are so central. And of all ascetic practices, *upavāsa* (fasting) is the one most favoured. It is not the case that all lay Jains engage in fasting, although every Jain of any level of religious seriousness fasts at least on the occasion of Paryuṣaṇa. Most men in their working years have too many other claims on their time and energy to engage in serious fasting, but even so some men fast on some fasting days.

Fasting, however, is primarily an expression of women's religiosity among Jains, and, if a woman's fasting is spiritually beneficial to her, it is also useful in both spiritual and material ways to her entire family. On the latter point, Reynell (1987; 1991) shows that a woman's religious activities, especially her austerities, attest to a family's moral soundness, which can have an impact on a family's creditworthiness. This is one of the reasons a woman's husband might finance a lavish public celebration of her successful completion of a major fast. But a woman's religious activities can also have a direct karmic effect on the well-being of other members of her family.

This raises an important issue. Because *karma* is an attribute of individual souls, not of collectivities, this means that the transfer of merit from the earner to someone else – an important idea in Buddhism – is seemingly precluded in Jainism. And yet the idea that others can benefit from one's own ascetic practices, ritual activities or pious donations is commonplace among ordinary Jains. Jain teachings do not seem to offer any clear-cut solution to this apparent paradox (see Cort, 2003).

To be efficacious, a fast must be undertaken as the result of a vow of reunciation made in the presence of a monk or nun who must give permission for the fast. A fast is believed to be religiously inefficacious if this step is omitted. Both asking and granting are done by means of standard verbal formulas. These formalities aside, there is plenty of room for individual choice (often instigated by monks or nuns) in the matter of how often or on what occasions or to what extent of severity one fasts (i.e., how long, with or without water, with two meals, one meal or no meals in given day).

Among Śvetāmbaras, the nearly universal fast is that undertaken during Paryuṣaṇa, and some individuals fast for the full eight days. Beyond that, there is a long list of possible occasions for fasting that individuals might or might not observe. An important fast is *āyambil*; it requires the faster to partake of only certain sour foods once a day, and women undertake it for the well-being of their husbands. Most people who perform this fast do so for the duration of a twice-yearly, nine-day festival known as Oli that occurs in the lunisolar months of Caitra (March/April) and Āśvin (September/October). Other fasting occasions include auspicious days associated with one of the Tīrthaṅkaras (such as Mahāvīra's birthday) and certain fortnightly days (mainly the fourteenth lunisolar day of both the waxing and waning halves of the lunisolar month, but also the eighth days and to some extent the eleventh and fifth days). Some individuals, usually housewives, undertake special fasts of a month or even longer. Successfully doing so is regarded as an extraordinary achievement deserving public recognition; the celebration takes the form of a congregational worship rite for friends, relatives and community followed by a shared meal.

The deceased grandmother of one of my Śvetāmbara friends was regarded as an upright woman of exemplary piety. As an example of orthopraxy, my friend described her fasting schedule as follows. She fasted for eight days on Paryuṣaṇa, of course, and also did the *āyambil* fast during Oli. She did a complete *upavāsa* (with or without water, depending on her stamina at the time) on every eighth and fourteenth day of the lunisolar fortnight, and fasted on some fifth and eleventh days as well. She also did one-meal or full fasts on Mondays and full-moon days on behalf of the Dādāgurus.

Śvetāmbara tradition also sanctions a variety of more complex and lengthy programmes of fasting. These are mainly undertaken by mendicants, but sometimes by laity under mendicant supervision. Often, lay Jains perform these fasts in groups in special camps where life is organised around the fasting and other observances such as temple worship and devotional singing in the evening. A simple but onerous example is the *siddhi tap* (Cort, 2001, p. 138). Performed by large groups of laity under the direction of a mendicant, it lasts for forty-three days and is designed to end on *samvatsarī* (i.e., at the end of Paryuṣaṇa). It consists of a one-day fast with only boiled water, followed by a one-day, two-meal fast; then comes a two-day fast with only boiled water, and then again a one-day, two-meal fast; and so on. This is continued until eight successive fasting days with water only are completed. The significance of eight is that this fasting programme helps one shed the eight types of *karma*.

These various ascetic performances are understood to give rise to spiritually salubrious thoughts and attitudes and also to work directly on the reduction of *karma*s (and in their very nature limit the influx of new *karma* while they are taking place). This, at any rate, is the soteriological interpretation that the mendicant establishment gives to fasting. The metaphor commonly employed is that of 'burning' *karma*s away. However, as Cort (2001, p. 138 ff.) points out, such ascetic performances are also associated with the accumulation of *puṇya* and the many kinds of worldly benefits (in this birth or another) to which *puṇya* leads. Cort notes that there is no inconsistency in this. By eliminating *karma*, ascetic practices push one ahead on the path to liberation, and because such practices primarily reduce bad *karma*s, they improve one's ratio of good to bad *karma*s, and thereby produce good worldly results (Cort, 2001, p. 141). Even so, in this as in much of Jain religious life, the accent is decisively on relinquishment of all worldly things (for which food is the obvious symbolic stand-in) and the ultimate attainment of escape from worldly bondage.

Also included within the category of fasting is the supreme fast called *santhāra* or *sallekhanā*. This is the ritual of voluntary death by means of ritualised self-starvation with which this chapter began. It is undertaken by laity as well as mendicants at a time of life when the body is in decline but before mind and will become compromised. It is not considered a form of suicide, and they say that one who takes the vow should hope for neither death nor life. The basic idea is that by reducing and finally eliminating food, and by doing so in a tranquil and detached frame of mind, one quells one's passions and sheds the encumbrance of the body in a state of total mental serenity.

Pilgrimage

Pilgrimage to sacred places is an extremely important dimension of the religious lives of lay and mendicant Jains alike. For a layperson, a formal pilgrimage is an opportunity to assume the temporary identity and lifestyle of a mendicant. However, to focus only on the path-of-liberation perspective on pilgrimage would be to miss some important aspects of the practice. Some pilgrimage destinations (such as Nakoda) are famous for their miraculous power to solve worldly problems and achieve worldly successes, and this is undoubtedly the motivating factor for many visitors. Also, in modern India, religious pilgrimage and tourism have to some degree merged, and recreation is at least part of the goal for many pilgrims.

While Jains go on pilgrimages in family groups and sometimes as individuals, an older pattern of group pilgrimage can still be seen today. Typically, a rich layman organises and finances such an event, and in so doing gains merit and community recognition for uprightness and piety. Nowadays, of course, such groups usually reach their destinations in buses, but if a mendicant accompanies the group then, of course, the pilgrims must move on foot.

Some pilgrimage sites are of subcontinental importance. Examples include places where Tīrthaṅkaras are said to have achieved liberation, and the most important of these is Sammet Shikhar in the state of Jharkhand. Here Parśvā and no fewer than nineteen other Tīrthaṅkaras gained liberation. Also important are Girnar in Gujarat, where Nemi (the twenty-second Tīrthaṅkara) obtained liberation, and Pavapuri in Bihar, where Mahāvīra achieved liberation. Of great importance to

Śvetāmbara Jains, in particular, is the vast complex at Mount Shatrunjaya in Gujarat. Tradition holds that this is where Ṛṣabha gave his first sermon, and his grandson, Paṇḍārīka, gained liberation.

Of international fame is the shrine at Shravana Belagola in the southern state of Karnataka. The main attraction here is the tenth-century statue of Bāhubali standing nearly sixty feet high. Bāhubali was the second of Ṛṣabha's one hundred sons. He fought a one-on-one duel with his elder brother Bharata (Ṛṣabha's first son and the first *cakravartin* [universal emperor] of our declining epoch and region or the world), but when he had the opportunity to strike the fatal blow he refused. Having become disillusioned with worldly things, Bāhubali renounced the world and stood motionless in meditation for a year. In the end, he achieved omniscience, and according to Digambara tradition he was the first to obtain liberation in our epoch and place. The statue at Shravana Belagola portrays Bāhubali standing in meditation (in what is called the *kāyotsarga* position) with creepers growing around his body. Every twelve years, a head-anointing ceremony is performed on the statue, a spectacular event that attracts thousands of Jains and others.

Leaving aside the great pilgrimage sites of subcontinental renown and importance, there are many lesser sites of mainly regional importance. For example, a pilgrimage centre in Rajasthan called Mahavirji (in Karauli District, and commonly written as 'Mahaveerji') is of great importance to Digambara Jains of the region. This is an example of the type of holy place known as an *atiśaya kṣetra* (place where a miraculous event occurred). In the case of Mahavirji, an image of Mahāvīra – the main image in the temple – was miraculously found underground when a cow let her milk down over the spot where it was buried. There are countless other such sites of regional and local renown in Rajasthan and elsewhere.

Chapter 6

A Moral Cosmos

My first foray into the study of Jainism took me to the Gujarati city of Ahmedabad. There I spent nearly two months taking notes on Jain ceremonies, mostly as I observed them in a single temple. While thus engaged, I met a Jain layman – a friend of an American friend – who was a doctor by profession but whose principal interest in life was Jainism and who was as serious a Jain as any I have met in lay circles. He did me the kindness of taking me under his wing and giving me a Jain catechism that extended over the entirety of my two-month stay in the city. In the course of this, he had little to say about the ceremonies I was watching. More important to him by far were the basic soteriological doctrines that have been described in this book, but not just these doctrines. He also spent a great deal of time on Jain cosmography, geography and biology. His insistence on the importance of these subjects was conveyed by the fact that he took the time and trouble to make for me some highly detailed drawings of the cosmos with its heavens and hells and also of the terrestrial surface on which we humans live.

I mention this because it points to what I thought then and still believe to be a very significant and possibly distinctive fact about Jainism. The Jains have created a large and very complex body of knowledge about the nature of the cosmos, the earth and living things. And, crucially, the Jains see this body of knowledge as a key part of their understanding of our creaturely condition and the possibility of liberation, which is why my friend took the trouble he did. It should be clearly understood that this knowledge is not arcane or obscure among Jains. Most Jains, lay or mendicant, are at least acquainted with its basics. Illustrated manuscripts contain detailed pictures of the cosmos and its structure, as do many of the books that lay Jains read and keep in their houses. Paintings and other representations of the cosmos are found in many temples. For these reasons, to understand Jainism one must take this body of knowledge into account.

The Jains' system of natural knowledge is not, however, comparable in its content and motivations to Western science. To put the matter somewhat differently, it is not – as some would say – bad science. Rather, it is a completely different kind of knowledge. It is a vision of the physical world, but it is also a conception of the world deeply shaped by a moral and soteriological perspective. It imputes certain values to natural phenomena, and the cosmos it describes is a moral cosmos, which is to say it is a cosmos to which the many and often onerous behavioural demands of Jainism are logical and morally appropriate responses. It is also a vision of extraordinary grandeur and imaginary power.

One more point before moving to the details. One will search in vain for a Jain theory of creation, for the Jain universe is uncreated; it always was and always will be. It is not changeless, but to the extent that it changes, which is not a universal feature of this cosmos, the alterations it undergoes are cyclical in character and the cycles are identical. To the extent that the Jains have a cyclical vision of cosmic time, it resembles the better-known Hindu scheme in which the universe is created, evolves and is destroyed in a final cataclysm, after which the cycle begins anew. In the Jain theory, however, there is no creation or final destruction, but instead an endless succession of epochs of moral and physical improvement and decline that succeed each other in a manner resembling a sine wave rather than the sawtooth Hindu pattern. For this reason, I employ the term 'cosmography' rather than 'cosmology' in what follows. Cosmology is generally understood to deal with cosmic origins, but the Jains have no such concept.

The Jain cosmos: time

As just mentioned, change is not a feature of the Jain cosmos as a whole; rather, cyclical movements occur only in certain small areas of the cosmos of which the part of the world we inhabit is one example. The cycle consists of periods of improvement and decline in an endless succession. An *utsarpiṇī* (ascending epoch) begins in a condition of moral and physical squalor. As things gradually improve, the moral and physical condition of human beings strengthens and nature becomes kinder and more bountiful; at the cycle's high point the world is a paradise. Then begins an *avasarpiṇī* (an epoch of gradual decline) in which all of the attainments of the previous epoch are gradually undone until the cycle bottoms out and a new ascending epoch begins. We are currently in a declining epoch. A complete cycle is known as a *kalpa.*

Each *utsarpiṇī* and *avasarpiṇī* epoch is subdivided into six separate eras, each defined according to how *suṣamā* (happy) or *duṣamā* (unhappy) it is. A *utsarpiṇī*, therefore, evolves from an era that is *duṣamā-duṣamā* (extremely unhappy), to *duṣamā* (unhappy), to *duṣamā-suṣamā* (more unhappy than happy), to *suṣamā-duṣamā* (more happy than unhappy), to *suṣamā* (happy), to *suṣamā-suṣamā* (extremely happy). An *avasarpiṇī* consists of the same eras in reverse. At the happiest end of the spectrum, human beings have immensely long lifetimes and are six miles tall. They experience neither conflict nor poverty. Effort is unknown, for their every desire is fulfilled automatically by wish-fulfilling trees. At the most unhappy extreme, humans live a maximum of twenty years, their height is only one and one-half feet, and their lives are spent in misery and want.

Precisely twenty-four Tīrthaṅkaras appear in each *utsarpiṇī* and *avasarpiṇī* epoch. Tīrthaṅkaras are born only in the second, third and fourth eras of a *utsarpiṇī* epoch and the third and fourth of an *avasarpiṇī* epoch. They do not appear during the extremes of happiness or unhappiness because too much happiness discourages the sense of urgency about liberation required to motivate ascetic practices, and too much unhappiness means that humans are too miserable to pursue liberation by engaging in such practices. Our current era in our *avasarpiṇī* is the fifth, and therefore no Tīrthaṅkaras will appear and nobody will achieve liberation (in our corner of the terrestrial world) until the next *utsarpiṇī* is well under way.

These cycles are immensely long. According to one version (J. L. Jaini's commentary in J. L. Jaini, 1974b, pp. 83, 89–90), a complete cycle (i.e., an *utsarpiṇī* and *avasarpiṇī*, or vice versa) lasts for twenty *koṛākoṛī sāgaropamās*. A *sāgaropamā* is a unit of time measurement and *koṛākoṛī* is a numerical expression equalling ten million squared, so a complete cycle requires 2×10^{15} *sāgaropamā*s for its completion. A *sāgaropamā* equals one *koṛākoṛī*, multiplied by ten, of time units called *addhāpalya*s, each of which in turn equals an uncountable (*asaṃkhyāta*, meaning finite but uncountable) number of *udhārapalya*s, each of which in turn equals an uncountable number of *vyavahārapalya*s. This last unit of time equals the amount of time required to empty a circular pit with a diameter and depth of one *yojana* (about eight miles) of fine lambs' hairs if hairs were removed at the rate of one every hundred years. I shall return to the meaning of these astoundingly vast numbers shortly.

In our own current epoch, an *avasarpiṇī*, we are in the fifth era. It began, as do all *avasarpiṇī*s, as a pre-cultural paradise. The *yugliya*s (humans of those days') were born exactly six months before their parents' deaths. They entered the world as mated sibling pairs and remained mated for the duration of their immensely long lives. This was a *bhogayuga* (age of enjoyment) because wish-fulfilling trees supplied everything the *yugliya*s needed without the necessity of toil. There was no thought of *saṃsāra* or liberation. Social and political organisation was unknown: there were no families, no clans, no kings, no social hierarchies.

So matters passed during the first, second and third eras, but conditions inevitably deteriorated. The wish-fulfilling trees gradually began to die out, which led to scarcity of food and other necessities. This was the beginning of *karmayuga*, our current age of endeavour. Scarcity, in turn, generated conflict and disorder, and people began to invest individuals of special abilities with authority. The *kulakara*s (leaders) instituted property rights and a system of punishments for malefactors, and in this way maintained social order as the former paradise dwindled away. The last of the *kulakara*s was named Nābhi, and he was the father of the first Tīrthaṅkara of our current epoch. Born in the city of Ayodhya, which the gods created as a capital for Nābhi, his name was Ṛṣabha.

While Jainism has no creator deity, Ṛṣabha, whose advent was at the end of the third era of our current *avasarpiṇī*, occupies this functional niche to some extent (P. S. Jaini, 1979, p. 288). The Jains maintain that he was the creator of culture, society and polity in our current *avasarpiṇī*. Having agreed to rule at the request of the *yugliya*s in order to deal with the ever-increasing disorders of the age, Ṛṣabha was also the first king. This is a Jain version of Thomas Hobbes's theory of the social contract. Ṛṣabha also taught humanity the basic arts of life, such as agriculture, fire-making and cooking, and he invented writing and mathematics. He also introduced the *varṇa* system (without the Brāhmaṇ class, which was later instituted by his son, Bharata). Ṛṣabha's was the first non-incestuous marriage (to Sunanda, whose male twin had died), which established the basic pattern of the family from that point forward. The Jains, it seems, have independently arrived at – and by their own route – the theory of Levi-Strauss and other modern anthropologists that pre-social humanity was humanity without the incest taboo. Ultimately, Ṛṣabha had two wives who bore him one hundred sons and two daughters.

In the end, Ṛṣabha abandoned the world. After this, he was the first of our *avasarpiṇi* to receive *dāna* in the form of sugar cane juice, an event already noted in connection with the annual festival (Akṣaya Tṛtīyā) that commemorates it. When he achieved omniscience and became the epoch's first Tīrthaṅkara, Ṛṣabha effected the transition between the third and fourth eras of our *avasarpiṇī*. He was not, however, the first to achieve liberation in our epoch. According to Śvetāmbara tradition, that distinction belongs to his mother, Marudevī (see P. S. Jaini, 2003). So impressed was she by the mere sight of his *samavasaraṇa* when she first approached it on the back of an elephant that she instantly rose through the *guṇasthāna*s and achieved liberation. This obviously cannot be a Digambara tradition, for they do not believe that liberation is possible from a female body. A Digambara tradition (mentioned earlier) maintains that Ṛṣabha's second son, Bāhubali, was the first to achieve liberation in our epoch.

During the ensuing fourth era came the remaining twenty-three Tīrthaṅkaras of our *avasarpiṇī*. Each propounded the same eternal truths, and each left behind the fourfold order of monks, nuns, laymen and laywomen that constitutes a Jain community. Each of these Tīrthaṅkaras had a particular history of past lifetimes leading up to his conception and birth as a Tīrthaṅkara-to-be, which is not only true of the twenty-four Tīrthaṅkaras of our *avasarpiṇī* but also presumably of all the infinite Tīrthaṅkaras that have ever been and ever will be. But from the spiritual standpoint, the final lifetime of every Tīrthaṅkara falls into a single eternal paradigm, that of the five *kalyāṇaka*s – the five events that define the career of all Tīrthaṅkaras: conception, birth, renunciation, omniscience and liberation. As has been described, the penultimate Tīrthaṅkara of our *avasarpiṇī* was Pārśva, known to be an actual historical figure, and the last was Mahāvīra. With Mahāvīra's liberation the fifth era, the current 'unhappy' era, began, and the window to liberation soon closed.

As the Jains conceive it, time itself, being beginningless and endless, presents a vast and essentially featureless landscape to the imagination. From the widest perspective, there is nothing to differentiate the endlessly repeating pulsations of sine-wave time. But the Jains have folded a narrative into the blank face of eternity that invests this timescape with moral and soteriological value and does so in a manner that makes sense of the world as experienced by men and women. Yes, the times are troubled, for this is the fifth age, but better things lie ahead, if very far ahead. And, for the spiritually awakened, the opportunity for final liberation will surely come around again (and indeed is currently available in other parts of the terrestrial world).

The narrative of our *avasarpiṇī* can thus be seen as mediating between eternity and the world as real men and women experience it, just as the eternality of the five *kalyāṇaka*s mediates the individuation of the careers of particular Tirthankaras and timelessness of the truths they teach. The eternal truths of Jainism tumble into the stream of time when the Tīrthaṅkaras reintroduce, as they do in an infinite series of advents, Jain teachings to the human world of messy history and perishable things.

The temporal cycles, however, are not a feature of the cosmos as a whole; rather, they are confined to a small zone. I turn now to the overall architecture of the cosmos as Jains understand it.

The moral cosmos: space

The Jain cosmos, the arena of rebirth, is a vertical structure, taller than it is wide, widest at the bottom, and vast in extent. Outside of it is merely non-cosmos. Its overall shape is not unlike the outline of a human figure standing with arms akimbo and legs spread, which indeed is how it is sometimes represented in artistic portrayals. This entire cosmos swarms with an 'infinite' (*ananta*, as opposed to *asaṃkhyāta*, 'uncountable') number of living things, which occupy its every cranny. Running from the very top to the bottom is a shaft known as the *trasa nāḍī*, so named because *trasa* (mobile beings with two or more senses) can live only within its boundaries. The simplest forms of life – the single-sensed *nigoda*s and element beings (to which we return in the next section of this chapter) – are found both within and outside the *trasa nāḍī* and occupy the comos to its very outer boundaries. At the very top of the cosmos is *siddha loka* or *siddha śilā* (a small zone shaped as an upward-facing crescent) where liberated souls abide eternally in omniscient bliss.

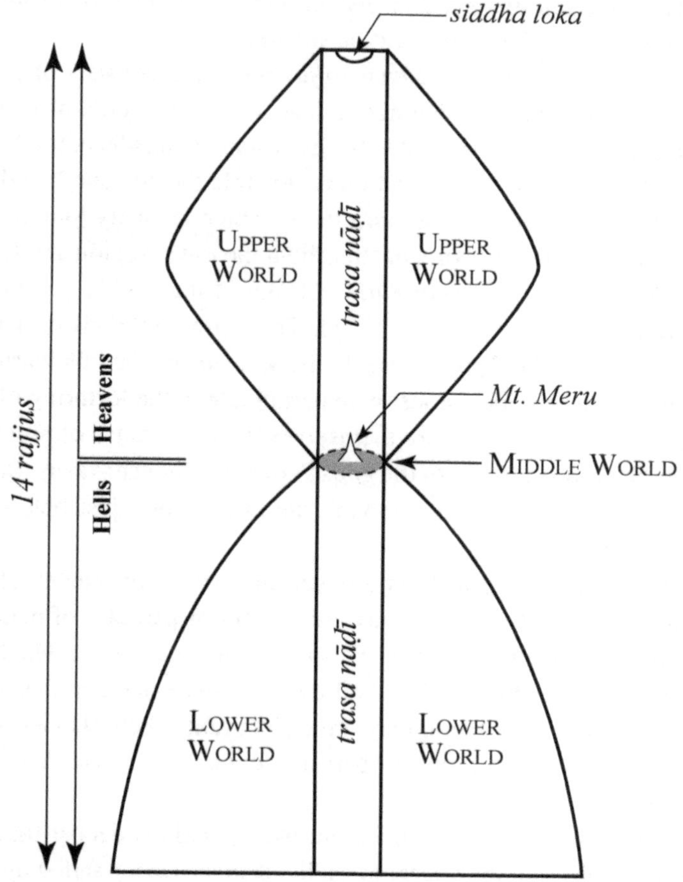

6.1 The Jain cosmos (greatly simplified, terrestrial disc tilted toward viewer).

Jain writings and traditions stress one point above all about the cosmos: its spatial vastness. As with time, the sense of size is communicated by means of huge numbers. The height of the entire cosmos is usually given in a unit of distance called the *rajju* (rope), and it is said to be fourteen *rajju*s high. According to one version, a *rajju* is equal to the distance covered by a god flying for six months at the rate of 2,057,152 *yojana*s per second (Caillat and Kumar, 1981, p. 20).

The basic subdivisions of the cosmos are three. Below the cosmos's waist is a hellish region known as *adho-loka* (lower world), above the waist is a heavenly region known as *ūrdhva-loka* (upper world). In the middle is a thin disc known as *madhya-loka* (middle world), which is where humanity is situated.

The lower *adho-loka* is seven *rajju*s deep and consists of seven layers of hell, each of a distinctive hue, where the souls of sinners suffer as a result of their misdeeds; the worse the misdeed, the lower the level of hell, the darker and more horrible the surround, and the longer the length of one's sojourn there, which is vastly long in any case. (Many of the following details about the lower region are drawn from Umāsvāti's *Tattvārtha Sūtra* 3.1–3.6 [Tatia, 1994, pp. 74–5]; also from Caillat and Kumar, 1981). There are also gods of a very inferior sort in the upper three hells, known as Paramādharmika (supremely unrighteous) gods, who participate in the torturing of the denizens of hell, and who are themselves in hell because of previous crimes. Not quite as low are the gods known as Bhavanavāsins (those who live in mansions); they inhabit the highest hell just below the base of Mount Meru.

Birth in any of the hells is asexual and occurs by means of the appearance of a fully grown body created out of particles of material afloat in the immediate vicinity. This is the same manner in which the gods are born (see below). The hell-dwellers share another characteristic of the gods; like them, they are clairvoyant. Their clairvoyance, however, enables them only to perceive things that cause them fear and pain.

The hells are extremely unpleasant places, and the deeper they are the more unpleasant they become. Hell's denizens are assailed by foul smells and are surrounded by streams of such noisome fluids as urine, faeces and mucus. They torment each other (although they are also tormented by the demonic Paramādharmikas), for they are constantly fighting among themselves with every weapon imaginable, including bare hands and teeth. They must undergo extremes of heat and cold, and they are afflicted with ravenous hunger and desperate thirst, neither of which is ever satisfied. Their suffering lasts until the karmic residues of previous misdeeds have been exhausted, at which point the now-unburdened sinners re-enter the round of births and deaths, as all unliberated beings must. It should not be thought that existence in hell is a 'punishment', at least not as that term is generally understood. It is simply an inevitable consequence of toxic karmic accumulations resulting automatically from certain kinds of behaviour.

Neither gods nor hellish beings can be reborn in hell because they lack the human capacity for moral choice. Hellish beings, moreover, cannot be reborn as gods; this is because they cannot engage in the pious observances and ascetic practices that rebirth in heaven requires. They can only be reborn as humans, animals or plants. The level of hell to which a being can be reborn is subject to limitations imposed by that being's moral capacities in its previous life. Thus, at the extremes, non-rational, five-sensed creatures cannot be reborn in any level lower than the topmost of the first level. Their non-rationality presumably rules out full understanding of the evil in evil deeds, and thus a lesser ability to do true evil. Reptiles with legs (which are considered rational and five-sensed) can be reborn in either of the upper two levels. Other classes of creatures have similar ranges – greater or lesser – of potential rebirth in hell. Male humans seem to have the greatest potential for evil, for they can be reborn in levels one through seven; that is, they can be reborn at the lowest level. It has to be said, though, that five-sensed aquatic animals (fish and crocodiles) can also end up in the seventh hell, which seems to mean that they, too, have a powerful capacity for doing evil. Interestingly, human females can be reborn only as low as level six.

The zone above the cosmic waist, seven *rajju*s high, is the realm of the gods. (Many of the following details about the heavens and gods are drawn from Umāsvāti's *Tattvārtha Sūtra* 4.1–4.42 [Tatia, 1994, pp. 97–117]; also from Caillat and Kumar, 1981). Although some lower gods live in the uppermost level of hell and in a lower zone of the middle region, and although planetary and stellar deities are situated above the surface of the terrestrial world, the *ūrdhva-loka* is the true domain of the gods and goddesses. It consists of ascending levels of heaven confined to the *trasa nāḍī*, each brighter, more subtle and the site of more intense enjoyments than the ones below. The deities who dwell in the heavens are known as Vaimānika because they travel in aerial vehicles known as *vimāna*s.

There is no suffering in the heavens, for the governing principle is that the deities of heaven are there to experience enjoyment. This happy condition is the result of the karmic residues of piety and good behaviour in the past; the better the record, the higher the heaven and the more subtle and intense the enjoyments therein. As has been noted about the sufferings in the hellish regions, the enjoyments of heaven are not to be understood as a divinely-conferred reward; they are merely the karmic consequences of past behaviour. Those very enjoyments of heaven are the reason why liberation is not possible from any of the regions of heaven. Enjoyment unleavened by suffering is unconducive to the mindset that gives rise to the desire to become liberated, and it cannot generate the fortitude that seeking liberation requires.

The Jains possess an extremely elaborate taxonomy of gods and goddesses. The fine points need not detain us, and indeed an accurate 'ethnography' of the Jain heavens is completely beyond the scope of a book of this sort, but some of the basics will be sketched out. All heavenly deities have special powers of locomotion and can travel to locations unreachable by human beings. All are clairvoyant. As are the denizens of hell, the gods and goddesses are born asexually by means of the drawing together of material particles from the vicinity; since the vicinity is heavenly, the nearby particles are of a nature far superior to those of the hellish zones. The lifespans of the gods are immensely long.

One of the most conspicuous features of the Jains' concept of heaven is the emphasis given to hierarchy, which is also true of the Jains' idea of hell (in some ways, a simple mirror-image of heaven). In the case of the gods and goddesses, the hierarchy is enormous in extent. There are four broad categories of deities ranked according to their relative altitudes. Leaving aside the demonic deities who mete out the many torments of hell, the lowest of the four categories is that of the Bhavanavāsins. They inhabit certain salubrious areas of the topmost hell. Above them are gods known as Vyantaras (forest deities) who live in the forests and mountains of a lower level of the terrestrial zone; they are close enough to the human realm to interact occasionally with humans. Then, high above the disc's surface are the solar, planetary and stellar deities, known as Jyotiṣas. Higher yet are the Vaimānikas, who occupy all the levels of heaven, properly speaking. Goddesses can be born only in the two lowest heavens, although they can pay temporary visits up to the eighth heaven.

Each of these categories also contains its own hierarchy in a manner reminiscent of the relationship between *varṇa* and *jāti* (Chapter 7). This is most notably true of the Vaimānika category. To speak in general terms, the fine-grained hierarchical distinctions among these deities are expressed by such characteristics as relative physical altitude (as with the four broad categories), colour, lustre, cognitive power, clairvoyant power and specialised function within the various heavenly realms (except for the very highest regions). These functions are mostly of the sort associated with pre-modern courts: kings, queens, ministers, attendants and so on.

The central Jain value of world renunciation is also reflected in the hierarchy, albeit only indirectly. The indirection arises from the premise that the point of a heavenly sojourn is enjoyment, which is inconsistent with ascetic practice. Thus, as a stand-in for ascetic values as such, the higher the divine rank, the fewer and weaker are the desires and the more subtle (and intense) the pleasures of the gods who dwell there. For example, the gods do enjoy sexual relations, but they do so in different ways. The lower categories of gods are accompanied by goddesses and gain pleasure from actual copulation with them. Gods at a higher level satisfy their sexual desires without actually uniting with goddesses. The lowest of the gods who satisfy their desires without union do so by touching goddesses who appear before them; at a higher level by simply looking at them; at a higher level yet by hearing them sing; and at a still higher level through sexual fantasy alone. The highest gods of all, and those whose enjoyments are the most intense, are those entirely devoid of sexual desire. These are gods whose quality of enjoyment is so sublimated as to nearly – but never quite – flip upside down into asceticism. So high are they that they are very close to liberation, and at the very pinnacle are gods that have but three or fewer births remaining before liberation.

As has been described, at the very top of the cosmos is the abode of liberated souls. Because the highest layers of heaven are just below this zone, it might be imagined that the ascending hierarchy of heavens represent a continuum of virtue, with the abode of the liberated at the end of the same road. This, however, would be quite erroneous, because there is a total disconnect between heaven, however high it may be, and the abode of liberated beings. This is because even the denizens of the highest heavens who are destined to attain liberation soon must return to the human condition before becoming liberated.

This brings us to an important point that must be understood. In the Jain view, heaven is neither permanent nor the highest goal of religious practice – and heaven is not even desirable when seen in the proper perspective. The gods and goddesses reap the karmic benefits of their piety and religious practices in previous lifetimes, and their fantastically long lives are totally devoted to enjoyment (albeit at graded levels of subtlety, as dramatised by the levels of sexual enjoyment discussed above). But the gods die in the end, however long their lifespans, and when they do they return to the round of birth and death, which indeed they never left, because heaven is part of *saṃsāra*. It is sometimes said that their felicity is uninterrupted until six months before their scheduled demise. At that point the flower garlands they have been wearing wilt and die. Then they are seized by remorse for the time they had wasted amidst the pleasures of heaven. It must never be forgotten that liberation from *saṃsāra* remains the all-encompassing aspiration of Jainism, and the only fully defensible one as well. Liberation alone is 'for keeps'; heaven and hell alike are simply gateways into further rebirth.

For a world-rejecting religious tradition to endure in the world as we find it requires that it secure a role within itself for a laity whose worldly endeavours make possible the support of a world-rejecting elite. Probably inevitably, this requirement gives rise to a subordinate domain of religious activity, seen as inferior from the perspective of the world-renouncer, that emphasises more worldly goals. This is the realm of the non-renouncing householder whose proximate goal is the attainment of auspiciousness and felicity in life and in lives to come. He or she achieves this aim by means of work in the world, pious practices and support of the mendicant community.

The status of the laity is not exactly second-rate, for it depends on the context in which the issue is considered. There is certainly plenty of room in the Jain way of life for worldly success (especially in business) and ritual celebrations of such success are very much part of the Jain way of life. But there is a limit. Asceticism is at the heart of many Jain practices for laity – most notably fasting – and it is commonly said that every Jain, whatever his or her current situation, should aspire to be a mendicant someday, even if that day is far in the future. No doubt such a wish is more a pious formality than truly felt for many, but it hovers over worldly life as a (perhaps only whispered) reproach.

Jain cosmography is a map of these ideas. Accretions of the wrong kind of *karma* (*pāpa,* demerit) will lead to trouble. That is what the hells are all about. Additions of the right kind of *karma* (*puṇya*, merit) will bring the enjoyments of heaven. But when seen from the perspective of liberation, the enjoyments of heaven are really no more desirable than the sufferings of hell, for *karma* – whether good or bad from the worldly perspective – is always bad when seen in the widest perspective, because its accumulation is the actual stuff of bondage. This is why Jains often say that bad *karma* is a chain of iron and good *karma* is a chain of gold, but both are chains.

The terrestrial disc

Human beings and all other forms of multi-sensed animal life as well as plants live only on a thin disc that separates the upper (heavenly) and lower (hellish) regions of the cosmos at its waist. Humans and animals cannot travel to either heavens or hells; indeed, humans cannot even travel to vast areas of the terrestrial disc.

The overall shape of the disc is circular. At its centre is a circular continent with a diameter of 100,000 *yojana*s (about 800,000 miles) known as Jambū Dvīpa (Black Plum Continent), and at its centre is a mountain called Meru, the axis of the world. Jambū Dvīpa is surrounded by a circular ocean – the Salt Ocean – which is twice as broad as Jambū Dvîpa is wide. The Salt Ocean, in turn is encircled by an atoll-like, circular continent named Dhātakīkhaṇḍa, which is twice as broad as the Salt Ocean and is itself enclosed by another ocean, again twice as broad as the continent it surrounds. This pattern of concentric continents and oceans is then continued to the edge of the disc. Each continent is twice as broad as the ocean before, and each ocean is twice as broad as the continent before, and the continents and oceans are *asaṃkhyāta* (uncountable).

The geography of Jambū Dvīpa is a very complex matter, but we need cover only its main features here. The entire continent is divided into seven zones separated by parallel mountain ranges: Bharata, Haimavata, Hari, Mahāvideha, Ramyaka, Hairaṇyaka and Airāvata. The smallest of these are Airāvata and Bharata, which are slender, D-shaped slices of territory at the northernmost and southernmost rims of the island respectively. They are tiny relative to the continent as a whole, but in human terms they are quite large, with a width of more than 500 *yojana*s (i.e., about 4,000 miles). Bharata (the Indian subcontinent) is the entirety of our world. On the island of Jambū Dvīpa, only Bharata and Airāvata are affected by the time cycles described earlier; at appropriate points in the cycles, they are *karmabhūmi*s (lands of endeavour) in which liberation is possible.

The largest zone in Jambu Dvīpa is Mahāvideha, which is a vast belt extending east and west across the middle of the island with Mount Meru at its centre. Mahāvideha is divided into four parts, two of which are *karmabhūmi*s where liberation is possible at all times because the time cycles do not operate there. The rest of the continent, including the other half of Mahāvideha, is *bhogabhūmi* (land of enjoyment), in which conditions are permanently unfavourable for the attainment (or even the quest for) liberation.

Human beings also exist on the atoll-like island of Dhātakīkhaṇḍa. The island has two Mount Merus of its own, and each is the centre of seven zones that are the same as those on Jambū Dvīpa – that is, there is a Bharata, an Airāvata and so on. The next island outward is named Puṣkaravara. This island is divided into concentric inner and outer rings separated by a range of mountains, and human beings can live only in the inner ring of the island. It, too, has two Merus, and each is at the centre of seven zones that are also the same as those on Jambū Dvīpa.

The time cycles apply only to the Bharatas and Airāvatas, and there is never more than one Tīrthaṅkara extant at any time in any of these zones. The Mahāvidehas are unaffected by the time cycles, and thus there are always Tīrthaṅkaras in the *karmabhūmi* areas of these zones, and at any given time there are always four per Mahāvideha. This means that even if there is no Tīrthaṅkara immediately available, as is the case in our Bharata at present, one can hope to be reborn in one of the Mahāvidehas where Tīrthaṅkaras are known always to exist. Although no human beings can exist beyond the inner ring of Puṣkaravara, animals and plants live there and the gods and goddesses freely travel in these areas. Also, some mendicants have acquired the power to reach these restricted regions.

Of the uncountable concentric islands beyond those where humans live, the most important is the eighth, known as Nandīśvara Dvīpa (Isle of Bliss). Nandīśvara Dvīpa is much celebrated in Jain tradition as a beautiful place where there are images of the four 'eternal Jinas', that is, Tīrthaṅkaras whose names recur in each cycle of time and in all of the regions of the world in which liberation is possible (i.e., *karmabhūmi*s) (Cort, 2010b, pp. 68–90). They are housed in eternal temples that are – as are the images held within them – perfect in beauty and immutability. This is a crystalline, out-of-the-stream-of-time realm of perfection that we, as beings in the stream of time, cannot experience directly. The gods resort to Nandīśvara Dvīpa three times a year to worship the eternal Tīrthaṅkara images there for eight days. The island is represented as an object of worship in various ways in Jain temples, and the gods' thrice-yearly pilgrimages there are commemorated (most notably by Digambaras) in thrice-yearly calendrical observances. The gods' worship of these icons is thus one of the charters for human image-worship. Eternal Tīrthaṅkara images and temples are also scattered elsewhere throughout the cosmos, though not in regions accessible to us.

Of particular importance are the Merus. The central Meru is at the centre of Jambū Dvīpa, but there are other Merus, as has been described. As are the heavens, hells and terrestrial zones beyond the inner ring of Puṣkaravara, all the Merus are inaccessible to human beings. Jambū's Meru extends upward into the lower heavenly regions and downward into the upper zones of hell, as presumably do the other Merus. Its height above the surface of Jambū Dvīpa is said to be 100,000 *yojana*s (about 800,000 miles) with an additional 1,000 *yojana*s (about 8,000 miles) extending below the surface of the disc. At its very peak are four pedestals facing the cardinal directions, and whenever an infant destined to become a Tīrthaṅkara is born in one of the regions below, the gods bring the infant to the appropriate pedestal where they give him his first bath. As has already been seen earlier in this book, this event plays a key charter role in Jain ritual culture, for the bathing of the Tīrthaṅkara-infant by the gods is the dominant paradigm for Jain rituals of worship.

Time and space

The Jains have made something of a specialty of large numbers deployed in the realms of time and space. Jainism is perhaps the only religion in which infinity is a core idea. And short of infinity, uncountabilities abound. These fantastic quantities apply to the cosmos as a whole, the terrestrial disc, Mount Meru, the temporal cycles, the lifespans of deities and hell-beings and much else. Large numbers are not unique to Jainism in the Indic world, but Jainism is different; not only are the quantities much larger than in other Indic traditions, but Jain writings also seem to linger over large numbers in a manner not seen elsewhere. (For a vivid account of the role of large numbers in Hindu tradition, see Zimmer, 1974.) The reason seems to be that such largeness reinforces the Jain insistence on the desperateness of our situation as living beings in bondage and the magnitude of the good fortune of those born in a human body and in contact with Jain teachings. One must seize that opportunity, the numbers say, with all of one's strength, for once one has been tossed back into this vast cosmic prison it will probably be an unimaginably lengthy wait, with much suffering, until one's chance comes again (Cort, 2010b, p. 68).

The world of life

Having seen the Birds' Hospital in Old Delhi, a well-known local landmark, a foreign newcomer to India might reasonably conclude that it represents an Indian outpost of the Western animal rights/animal liberation movement. But our visitor would be mistaken. The Birds' Hospital has nothing whatsoever to do with the West; it is neither a creation of Western animal rights activists nor was it inspired in any way by Western animal-rights advocacy. It is, rather, an institution established and supported by the Jains, and it stands as a testimony to the likelihood that no human community has thought more deeply about the relationship between humans and other species than the Jains, and more strongly objected to harm inflicted on living things.

As described in Chapter 3, the Jains insist that all forms of life are fundamentally alike in respect to the souls embodied in them. Every living thing has the capacity to feel pain and to suffer, and, as did Jeremy Bentham in the West, the Jains have regarded the capacity of non-human beings to suffer as imposing an obligation on humans to refrain from causing them harm. It is true that the Jains did not couple this thinking with an explicit concept of animal 'rights', because this idea was not part of their intellectual heritage, but the moral implication of their position is more or less the same (Babb, 2011). There is, however, a big difference as well. To the best of my understanding, it would never have occurred to Bentham to extend the umbrella of protection to *all* forms of life, including plants and even microbes. And yet this is precisely what the Jains have done, which in turn has had a large impact on the way the Jains visualise their social order.

Let me start with what seems to be an inescapable social implication of the Jain ethic of non-harm. If there is a consensus within a religious community that all forms of life, no matter how humble, can suffer and must be shielded to the fullest extent possible from hurt, then that community will have to institutionalise a division between first- and second-class religious citizens. The reason is simple: the continuation of life itself, even for the strictest of vegetarians, requires the harming of some other living things. The most basic activities of daily life – food preparation is a good example – involve harming and killing vast numbers of living entities. Therefore, if anyone is to be supported in a life that maximally protects all forms of life, then someone will have to compromise and adopt a lifestyle in which a greater degree of harm inflicted is considered tolerable. Virtue, in other words, will have to be socially organised.

The Jains meet this requirement by means of their basic social divide between mendicants and laity. Of course, respect for life is not the only value served by this division; mendicant asceticism is also made possible by the (relatively) non-ascetic lay majority. This much is common to all monastic traditions. But among the Jains the division is greatly accentuated by the fact that Jain mendicants are required to live in a manner that protects even the smallest and lowest forms of life. Because this is inconsistent with obtaining even the most basic requirements of daily life, the mendicants must rely for their subsistence needs on a non-elite stratum whose members are willing to inflict some degree of pain and death on some forms of life. Furthermore, these two classes must come into daily contact with each other to ensure the full meeting of the subsistence needs of the mendicants.

But the matter does not end there. Although Jains must of necessity divide the spiritual labour between partial and nearly complete fulfilment of the commitment to non-harming, there remains a range of worldly activities involving harm to human and non-human life that a viable society needs to have done and that are too egregious even for a Jain laity. By society I mean a complex society in which Jains are one component; quite obviously, there can be no Jains in a hunting and gathering social milieu. In complex societies there is a range of necessary occupations that require the inflicting of pain or death on living things – pest control, the punishment of criminals, livestock management and many others – that Jains are forbidden or, in less flagrant examples, at least discouraged from entering (soldiering is a special case, to which we return in the next chapter).

Given the apparent necessity of these forms of violence within the wider social order in which Jains find themselves, yet another social distinction must be added to the division between mendicants and laity. This is a moral divide between the entire Jain community and the rest of the human world, a world in which violence of a sort not possible for Jains is an accepted necessity. From the strict standpoint of Jain teachings, this wider community lies outside the circle of true moral understanding (although non-Jain vegetarian communities can be said to occupy an intermediate position).

These considerations point to a crucial practical implication. If it is to be consistent with a viable way of life in the world as we know it, a universal ethic of non-harm requires not only morally-grounded social distinctions but also moral distinctions between different kinds of living things. That is, in order that some protection can be extended to all forms of life in a manner consistent with the subsistence needs of human beings, some forms of life are going to have to be more protected than others. But which forms of life and on what criteria? In order to answer such questions in an intellectually defensible way, there needs to be a system of biological knowledge. The Jains have produced just such a system. It is an impressive intellectual construct, highly complex and with many fascinating byways, and a full accounting of it cannot be pursued here. But here are some of its basics.

Living things

Jain teachings portray the cosmos as swarming with forms of life that are born, live and die in its every cranny (on all these points, see Sikdar, 1974; Javeri and Kumar, 2008; Tatia, 1994, pp. 41–54; a recasting of the system in modern terms can be found in Bothara, 2004). As is true of Western scientific biology, taxonomy is at the heart of Jain biology. However, the Jain system of classification differs from the Linnaean system, not just in the categories it employs but also in the very basis of classification itself. It is based on structural criteria, as is the Linnaean system, but the conceptual context in which these criteria are deployed is soteriological and moral. The question lying at the heart of the Jain system is that of the implications of a living thing's nature for the way a morally enlightened person should interact with it. From the perspective of Western biology, the Jain system is thus not a system of biological knowledge as such; rather, in parallel with 'moral cosmography', and to use a helpful expression coined by McKim Marriott (1976), it is a system of 'bio-moral' knowledge.

In accord with an idea widespread in the Indic world, the Jains believe that there are 8.4 million distinct forms of life in the world. This number is a mere convention; what seems to be meant is that the total is very, very large. The term generally used for a particular kind of living thing in this context is *yoni*. This word carries the basic meaning of womb or female genitalia, and when applied in this context it quite logically points to birth (even though many forms of life are not womb-born) as the crucial connecting point between episodes of life in the world careers of souls. P. S. Jaini (1980, p. 130) renders *yoni*, when used in this context, as 'birth situation' and notes that these are the 'species' of the Jain world of life.

The *yoni* concept, however, maps only loosely on to another and far more important classificatory scheme – one that relates directly to Jain soteriology and its implications for the treatment of non-human life. This second taxonomic level, which we might consider Jain biology's generic level, assigns all living things to four broad categories called *gatis*, what P. S. Jaini (1980, p. 125) calls 'destinies'. The four *gatis* are (1) humans, (2) deities, (3) dwellers in hell and (4) plants and animals. This scheme is basic to Jain teachings and is shared by all branches and sects.

The term *gati* carries the basic meaning of 'movement' to from one state or condition to another, with the emphasis on the destination, suggesting a spatial dimension to the concept, and in fact each *gati* occupies, more or less, its own place in the cosmos organised along a partly vertical axis of worth (P. S. Jaini, 1980, p. 125). It is, one might say, a 'spatio-moral' as well as a 'bio-moral' system in the sense that the altitude at which one ends up depends on the moral nature of one's karmic legacy.

As has been described, the Jain cosmos consists of three basic parts: a multilayered heaven above; a multilayered hell below; and a thin disc in between (leaving aside the abode of the liberated at the very top). The higher gods dwell in the heavens, and their polar-alters, the hell-beings, inhabit the hells. Living on the intermediary disc are humans and animals and plants, with humans confined to its central zones. While the vertical dimension of the cosmos is continuum of value (higher is better), there is an important exception. As has been seen, liberation, which is the highest value in Jain teachings, is possible only in a human body. Considered in this perspective, the human *gati* is the highest of the four.

The *gati* of animals and plants, called the *tiryañca gati*, is fundamentally different from the other three. The *gatis* of human, deities and hell-beings are relatively undifferentiated; that is, all members of each one are beings of the same basic type, although they differ in many other respects. By contrast, the *tiryañca gati* is variegated to an extreme degree, and it is here that the *gati* and *yoni* systems truly intersect; in effect, the *tiryañca gati* is the home of the 8.4 million *yonis*.

The main principle organising Jain biology within the *tiryañca gati* is structural and focused on the senses; it classifies living things according to the number of senses they possess. Deities, hell-beings and humans all possess five senses, but the *tiryañca gati* contains the entire sensory range of one to five senses and comprises a hierarchy with five-sensed beings at the top. Gradual upward progress is possible within this hierarchy, for a soul can advance up the ladder of plant and animal life, ultimately to acquire a human body. Such progress is not possible within the *gati*s of deities and hell-dwellers. From this standpoint, a sojourn in heaven or hell is a spiritual 'time out'.

At the very bottom of the *tiryañca gati* are living things possessing only one sense, that of touch, and incapable of movement. They are of several types. The simplest of all forms of life, and lowest on the entire ladder of living things, is known as *nigoda* (on which see P. S. Jaini, 1980; Sikdar, 1974, pp. 95–9). The *nigoda*s are invisibly tiny entities, infinite in number, and are found throughout the cosmos, even in the bodies of other forms of life (including humans, but not in the bodies of deities or hell-beings). They spend their brief lives in tiny clusters, each containing an infinite number of souls that take birth and die together. For some souls, the status of *nigoda* is the beginning of an upward career, but it is also quite possible to fall back into this humblest of all living states, even from a human body. Also very low on the ladder of life are the one-sensed living things embodied as earth, air, fire and water, which, as described earlier, means that these elements are actually alive.

At a somewhat higher level, but also one-sensed and immobile, are plants properly speaking; they, in turn, are divided into two classes: *pratyeka śarīra* (plants in which a single soul inhabits a single body) and *sādhāraṇa śarīra* (plants with multiple souls in a single body). Prominent among the latter are root crops such as potatoes, carrots, onions and garlic, which Jains are supposed to avoid precisely because of this multiplicity of souls. Plants are found only on the terrestrial disc, unlike the ubiquitous *nigoda*s and element beings.

To move from one-sensed forms of life to those with two or more senses is also to shift from immobility to mobility (although some sources consider the fire- and air-embodied organisms to be mobile). All *trasa* (multi-sensed mobile beings) are confined to the chimney-like *trasa nāḍī* running from the top of the cosmos to the bottom; it encompasses the heavens, the hells and the entirety of the terrestrial disc.

The sense of touch is universal to the world of life – even the lowly *nigoda*s and element beings have this sense – and this is the base to which other senses are added. The additional four senses are those of taste, smell, colour (i.e., sight) and sound; and the organs of sense are skin (for touch), tongue, nose, eye and ear. The most primitive of mobile beings (such as worms and molluscs) have the senses of touch and taste. At a slightly higher level are small creatures (ants, centipedes, fleas, termites) that add to these the sense of smell. At a higher level yet are certain other small creatures (flies, wasps, mosquitoes, butterflies, scorpions) that also possess sight. The largest and highest of living things, apparently mostly vertebrates and including humans, deities and denizens of hell, have the fifth sense of hearing.

The five-sensed animals have a special status (P. S. Jaini, 1987; Wiley, 2006b). As are humans and deities, they are present at the Tīrthaṅkara's preaching assembly. They, too, are rational (if only in a limited way) and have the capacity to understand and benefit at some level from the Tīrthaṅkara's teachings, and even to develop spiritual insight. Under the right conditions, they are capable of moral choice and can choose to practise non-harm (or decide not to). They cannot, however, rid themselves of their karmic burdens directly, which is possible only in a human body.

The key point is that – however many senses they possess – all living things engage in sensory interaction with the world, even if only through the sense of touch. This means that all living things have the capacity to feel pain and to suffer. The simpler forms of life cannot hurt in a way that is self-aware, but even if they do not know what is causing their suffering or how much they are feeling, they do feel pain (Wiley, 2006b, p. 251).

We may wonder why the sensory principle is sequestered in one *gati* and does not organise the *gati* system as a whole. The answer seems to lie in a functional difference having to do with human engagement with these two very different systems. One system – the one internal to the *tiryañca gati* – relates to how we as humans should deal with other forms of life, while the other system – the *gati* system – has largely to do with the results of our behaviour.

The *tiryañca gati* confronts us not only with a vast range of possible rebirths but also with a range of moral choices that influence our future rebirths. The issue of causing or not causing suffering cannot arise with respect to deities or denizens of hell because these beings are beyond the reach of human contact – good or bad. What remains are other human beings and plants and animals. Humans are covered by a blanket injunction against inflicting pain or causing death, a matter that requires only reiteration and emphasis, not elaboration. In the large *tiryañca gati*, however, the sheer variety of contexts in which these beings are encountered – including and especially dietary contexts — necessitates more complex distinctions. Significantly, these divisions are largely based on the number of senses these forms of life possess, that is on their capacity to suffer.

The prohibition of harm to the larger and more complex creatures applies to all Jains, leaving non-Jains to do as they will (as indeed the welfare of society at large may require). But in deference to the practical issues of subsistence and governance, lay Jains are released from the commitment to refrain from harming the simplest life forms. This commitment, however, is retained by the mendicant community as one of their most notable obligations. There are gradations in between: for example, plants containing multiple souls (such as root crops) should not be eaten, and yet this is considered a much less serious infraction than eating meat. In contrast, and in consistency with its emphasis on transition and destination, the *gati* system is fundamentally about the future consequences of the things we do, and the categories that count from this standpoint are three: heaven for the virtuous; hell for the non-virtuous; and the human *gati* as the most desirable of all (from a soteriological standpoint), for liberation is possible only in a human body.

Lastly, it must be noted that Jain teachings posit an affinity of sorts between each and every living thing, from the gods and goddesses in their heavens to the most humble microbes. It must not be imagined, however, that this has any relation to the Western idea of the 'web of life', for the idea that shared ancestry might be a means of understanding relationships between forms of life plays no role whatsoever in Jain biology. To some extent its spirit is similar to that of the Linnaean taxonomic system before it became impregnated with Darwinian ideas, but here, too, there is a crucial difference. Jain biology is not just a system of classification, but also a system of classification that serves normative and soteriological goals. Nor does it bear any resemblance to the Western concept of ecological interdependence uniting the world of life into a single system. Jain bio-morality is largely uninterested in the world of living bodies as such, which, in fact, it radically devalues. Rather, the Jain idea of the affinity of living things focuses on the empathetic fellow feeling that arises from a sense of identity with others, and this identity is that of the soul and has nothing to do with the physical body.

Chapter 7

Social Context

Jainism was once, long ago, a proselytising religion. This is no longer so, and has not been the case for centuries. In this I am ignoring the seventeenth- and eighteenth-century conversions of image-worshippers to the Sthānakavāsī and Śvetāmbara Terāpanthī sects as conversions to Jainism, and current efforts to convert some of India's tribal peoples are a marginal phenomenon (see Dundas, 2003, pp. 142–4 on this issue). Jain teachings, therefore, continue to exist in the world because they are transmitted within enduring social groups, not because of the conversion of non-believers. In referring to such groups, I do not mean the idealised Jain social order as projected in texts, but the actual social groups formed by Jains. And if our task is that of understanding Jainism in any comprehensive sense, we must know something about the nature of these groups and the ways in which they fit into broader social contexts in India. This is the topic of the present chapter.

If we begin with a bird's eye view, we first notice that Jains are a tiny minority of India's population. Because Jains sometimes identify as Hindus, the total number is certainly larger than the 4.2 million recorded by the 2011 Census of India. Still, Jains are but a sliver of the whole. The second thing we spot is that Jains tend to occupy certain specific social and economic niches in Indian society. Most Jains – not all, but most, particularly in India's North – belong to communities whose traditional occupations are trade and moneylending. These communities are usually called 'castes' in English, and they are among the most important institutional contexts in which Jainism is socially transmitted. How do most Jains become Jains? They do so by means of birth in Jain families, and Jain families belong to castes or sections of castes that bear a strong social identity as 'Jain'. Therefore, to understand the social context of Jainism, it is necessary first to learn something about the nature of caste in India.

Caste basics

Any discussion of caste in India must begin with a system of social classes that are not in themselves castes but that must be understood if the caste system as a whole is to be comprehended. This is the *varṇa* system that was touched on in Chapter 2. The *varṇa*s are a system of four-ranked, hereditary classes that was first described in a late hymn of the *Ṛg Veda* (10.90). The system was later elaborated in the Hindu law books, and to this day *varṇa* continues to provide a vocabulary and conceptual framework with which Hindus describe and understand their traditional social order in the most general terms.

The four *varṇa*s together form an ideal social order based on a simple division of labour between classes. At the apex (albeit a contested apex) are the Brāhmaṇs, the priests and teachers of the system. Next in status are the Kṣatriyas, the ruling and warrior class. Third in status is the Vaiśya class; originally, this comprised ordinary agriculturalists, but with the evolution of urban society the Vaiśya label came to be applied mainly to merchant groups. In theory, males of these three classes undergo a ritual of initiation that is viewed as a second birth; accordingly, these three categories are called *dvija* (twice-born). At the system's bottom is the class known as Śūdra; they are not initiated, and the classical law books describe their occupational function as that of 'serving the twice-born'. Excluded from the system altogether are those called Caṇḍāla in ancient times; they were the prototype of the groups that came to be known as 'untouchable'.

The four *varṇa*s never existed as actual groups of any sort, nor, as has just been stated, are they the same things as castes, although the *varṇa* labels are sometimes used as if they referred to castes. Rather, they are categories that function as a pan-India conceptual system for the classification of castes. But what then are castes?

Social scientists who study this subject usually reserve the term 'caste' for a type of social entity known in Hindi as *jāti*. This term has a range of related meanings, but in the present context it refers to named, in-marrying social categories that are usually associated with a particular region and traditional occupation. For example, a *jāti* found in wide areas of northern and central India is called Telī, a name derived from *tel* ('oil' in Hindi), and they are the oil pressers of this region. Some *jāti*s are named for their supposed place of origin. The Agravāls are a very large trading caste of northern India, and are so called because they trace their origin to the ancient city of Agroha in what is now the state of Haryana. In each of India's linguistic regions will be found an array of such social entities, and some of them extend spatially beyond regional frontiers.

While there is no need to dwell on the technicalities of this subject, it is, nonetheless, important that readers understand that *jāti*s are not actual groups; they never congregate in a single place, and they cannot act in a concerted fashion (although in modern times bureaucratic organisations called 'caste associations' often act on their behalf). They are best understood as regional social identities (broadly speaking), often linked to traditional occupations, under which much more localised *jāti*-segments categorise themselves and are classified by others. Among their other functions, caste identities play a key role in restricting marriage choice; that is, whatever other considerations and rules there might be, a bride and groom should come from the same *jāti*.

Caste identities carry varying degrees of social honour, which means that the castes coexisting in subregional and local communities typically form hierarchies. Rank in such hierarchies is partly (but only partly) conceptualised in terms of purity (high status) and impurity (low status), often linked to culturally-based judgements about the 'dirtiness' or lack of the same of traditional occupations. These rankings are, to some extent, replicated in urban situations, but urban society is a relatively poor medium for the maintenance of strict hierarchies.

The *varṇa* system classifies *jāti*s within a framework that has subcontinental meaning: for example, in the state of Rajasthan there is a *jāti* (i.e., caste) known as Dādhīc Brāhmaṇ. Members of this caste marry only each other, worship the same caste goddess and take pride in their caste and the achievements of its members. However, it is not the only Brāhmaṇ caste in Rajasthan, for there are several others. These various Brāhmaṇ *jāti*s are all accorded the status of Brāhmaṇs because they so distinguish themselves and are so identified by others. Other regions have their own Brāhmaṇ *jāti*s, such as the Iyer and Iyengar Brāhmaṇs of India's deep South. The *varṇa* category 'Brāhmaṇ' provides a nominal identity, associated formally (if not in fact) with priestcraft and teaching, that transcends regional boundaries.

Jains and caste

With this background in mind, let us turn to the Jain understanding of caste. A crucial fact here is that Jain scriptures are, as far as I know, silent on the subject of *jāti*. This is not true of Jain tradition, as will be seen, but the social fissures separating *jāti*s are essentially invisible from the lofty perspective of Jainism's most sacred writings. The *varṇa* system, however, is another matter.

The Jain understanding of the *varṇa* system is very different from that of the Hindus. As described in Chapter 2, hymn 10.90 of the *Ṛg Veda* legitimises the *varṇa* hierarchy by linking it to the creation of the cosmos. If, as the hymn suggests, the *varṇa* categories are simply part of the natural ordering of things in the cosmos, then their legitimacy is hard to question. But for the Jains this cannot be so for the simple reason that the Jains maintain the world was never created. The *varṇa* system, therefore, has to have been a human invention, and its creator was Ṛṣabha, the first king and Tīrthaṅkara of our declining epoch and place in the world. This is the source of its legitimacy.

The narrative in which this idea is embedded is the quasi-Hobbesian theory of the origination of civilisation that was encountered in Chapter 6. During the previous eras, there had been no need for social or political institutions at all because there was no competition among the humans then living, due to the wish-fulfilling trees. The first step was taken when Ṛṣabha became the first king in order to quell the disorders that had broken out in the population. This was the origin of the Kṣatriya *varṇa*. Then the Vaiśya and Śūdra *varṇa*s emerged in order to accommodate the various occupations that Ṛṣabha created.

The Brāhmaṇ *varṇa*, however, was a special case. Jains have always had an adversarial relationship with Brāhmaṇs; they rejected the Vedic heritage to which Brāhmaṇs are central, and reserved a special distaste for the Vedic sacrifice, which often involved the immolation of animals. An oft-repeated charge heard from the Jains is that, while the Brāhmaṇs claim that the sacrifice conveys its animal victims directly to heaven, they actually perform such rites only to eat the meat.

Ṛṣabha himself did not create the Brāhmaṇ varṇa; it was his son and successor, Bharata, who did. Bharata, who was the first *cakravartin* of our declining epoch, decided to test the understanding and commitment of the lay Jains. He invited them to a festival, and then had flowers and sprouting grain spread out on the floor of a courtyard they would have to cross on their way into the palace. Some refused to violate *ahiṃsā* by treading on these living things, and Bharata invited these pious individuals to accept one or more of the *pratimā*s. He also ceremonially invested them with sacred threads (the rite that constitutes the second birth of males of the top three *varṇa*s in Hindu society) and gave them the title 'twice-born', saying that they had been reborn as 'children of the Jina'.

This narrative expresses a general outlook on human social institutions that stands in dramatic contrast to the Vedic/Hindu theory of *varṇa*. In what is obviously a riposte to what was seen as the arrogance of Brāhmaṇs – who considered themselves to be godlike because of their Brāhmaṇ descent – the narrative presents a very different view. In the Vedic/Hindu view, the hierarchy of *varṇa* actually carries sacred value; that is, the *varṇa* categories are linked with differing inborn codes for behaviour that are seen as divinely created and inherent to the very nature of different categories of human beings (Marriott, 1976). The Ṛṣabha narrative argues that – to use a modern locution – these differences are cultural, not natural, and have been imposed by human agency in response to human needs. All of this is quite consistent with the Jain view that – the fourfold order of Jain society aside – social arrangements as such do not bear sacred values. And as for the Brāhmaṇs, the Jain view was that true Brāhmaṇhood is a matter of knowledge and right conduct, not birth.

This is not to say that *varṇa* identity is without importance in the Jain world. While Jains have found their way into a great range of occupations, and this despite the constraints that *ahiṃsā* exercises on occupational choice, on the whole – and especially in the North – commerce has been the principal field in which Jains have sought a livelihood. On this criterion, most Jains are categorised as Vaiśyas, and so consider themselves. Moreover, *varṇa* plays a key role in the origin myths of several important Jain castes, a subject to which I return shortly. But, on the whole, Vaiśya identity is of minor importance among Jains. As for their status as traders, the word 'Baniyā' (a Hindi term for trading communities, Hindu or Jain) is heard much more frequently, even in self-description, despite its negative connotations of miserliness and hard dealing. In any case, what matters most in actual social life is *jāti*, not *varṇa*.

The importance of *jāti* in the social organisation of Jain communities might come as a surprise to a reader of what has been described thus far about Jains and caste. It is true, nonetheless, that if Jain doctrine does not sanctify caste, which it does not, caste definitely plays a key role in organising Jain communities as they exist on the ground (see Ellis, 1991 for a good description of Jain castes *in situ*). In doing so, caste intersects with other identities in several permutations. On one side is religious identity: depending on the context, Jains identify as Jains, as Śvetāmbara or Digambara Jains; or as members of one of the subsects within the Śvetāmbara or Digambara fold. On the other side is caste identity itself. Some castes are entirely Jain, others are only partly so; some are Digambara, others Śvetāmbara, and so on. The upshot is a complex social map of which the following is a quick overview.

Caste and Jains: The South

The deepest ethnographic fissure in India's Jain world is between the North and the South; not only is there a sectarian divide but also major cultural differences exist between the Jains of the two regions. As seen in Chapter 2, India's South is where Digambara Jainism found its home, although there are significant numbers of Digambaras in the North as well. There is a scattering of Śvetāmbaras in the South, but mostly they are recent migrants from the North who came there in response to business opportunities. The Jains of North India belong to castes that are considered trading and banking castes, and this is true of the Digambaras as well as Śvetāmbaras. The Digambaras of the South tend to be farmers or small shopkeepers and artisans, although a professional class has grown over the last century or so who have taken the lead in reform movements (Carrithers, 1991).

Because of their prominence in trade, both Śvetāmbaras and northern Digambaras are far wealthier than the Digambaras of the South.

The Jain castes of the South are few in number and are exclusively Digambara Jain (Sangave, 1980, pp. 95–101). The *bhaṭṭāraka*s, the domesticated Digambara functionaries who were mentioned in Chapter 2, once functioned as their caste gurus and rulers, and also served as mediators between their mostly impoverished rural followers and the political authorities of the region. One author suggests that some of these agrarian Digambara castes were actually created by the *bhaṭṭāraka*s in order to have stable followings (ibid., pp. 96–7). This is an unlikely account of caste formation, but it does illustrate the extent to which the castes in question and Digambara Jainism are interlinked in the South. A more plausible scenario is that the absence of ecclesiastical institutions transcending caste and locality (such as the *gaccha*s in the North) created a situation in which religious authority settled easily into the social shell of pre-existing castes. The power of the *bhaṭṭāraka*s is much diminished from what it once was, but something of a revival of the institution appears to be under way currently.

A notable feature of the South Indian Jain world is the development of hereditary classes and/or castes of Jain priests. One such group is the Upādhyāya caste, whose members serve as priests in Jain temples (Sangave, 1980, pp. 98–9). Educated laymen are apparently allowed to join this caste, but once in it they must obey the rules of caste endogamy (that is, they can take brides only from, or send daughters to, other members of the caste). They not only perform temple ceremonies, but also specialise in officiating at rites of passage and other occasional rituals for local families (Carrithers, 1991, p. 277). Also in the South is a strictly hereditary class of Jain Brāhmaṇ priests, who are knowledgeable in Jain scripture (as opposed to Hindu scripture, with which normal Brāhmaṇ priests are supposed to be familiar). They are apparently attached to temples as priests, and also officiate over a broad range of domestic ceremonies (Sangave, 1980, pp. 100–1). While their Brāhmaṇical credentials might seem problematic to Hindus, their Jain followers regard them as the descendants of the Brāhmaṇs long ago certified by Bharata. The Jains say that the Hindu Brāhmaṇs are the descendants of those who have 'fallen away from the true path' (P. S.

Jaini, 1979, p. 291). As already noted in Chapter 5, the priestly role is essentially absent among Jains in the North.

Jains and caste: North India

In contrast to the South, North Indian Jains, whether Śvetāmbara or Digambara, belong to castes specifically associated with trade and moneylending, which, in fact, is the normal occupation of most North Indian Jains. Also, some of these castes have both Jain and non-Jain membership. As one might expect of castes specialising in trade, these castes tend to be quite widespread geographically. As an example, members of the Agravāl caste – a trading caste with a significant Digambara Jain membership – can be found virtually everywhere in India, although they originate in the North. An additional point of great importance is that, with minor exceptions, the Digambara and Śvetāmbara Jains of the North belong to different castes, which is one of the reasons for the persistence of the social chasm between the two branches. The requirement of caste endogamy means that the kinship networks engendered by marriage cannot (under normal circumstances) extend across the sectarian divide, and kinship ties constitute a basic framework for some of the most important social relationships in Indian society.

I now turn to three North Indian Jain (or partly Jain) castes and their origin narratives. (For more extended details and analysis, see Babb, 2004.) These narratives will provide some idea of the way Jainism and caste are intertwined at the ground level. As noted earlier, there is no recognition of the social reality of caste in Jain scripture. Nevertheless, Jainism is woven into the very fabric of social identity at the level of caste, and origin narratives are the principal expression of this fusion of the religious and the social. However, these are myth-histories and must not be read as accounts of how these castes were actually formed, which is beyond the reach of certain knowledge. The examples are drawn from the Rajasthani city of Jaipur and its hinterlands, which is the area of India I know best.

Khaṇḍelvāl Jains

The Khaṇḍelvāl Jain caste has spread over the entirety of North India, and such cities as Mumbai, Kolkata, Indore and Jaipur are considered Khaṇḍelvāl Jain centres (Kāslīvāl, 1989, p. 63), but Rajasthan is the caste's geographic homeland. The caste bears the name 'Khaṇḍelvāl' because its members maintain that it originated in a place called Khandela. Although Khandela is little more than a small town in north-eastern Rajasthan today, in legend it was once a large and important city. 'Khaṇḍelvāl' is a contraction of *khaṇḍela vālā* (of Khandela), and at least four castes, among them the Khaṇḍelvāl Jains, bear this name. The Khaṇḍelvāl Jains are also called 'Sarāvgī', which is a version of the term Śrāvaka (i.e., Jain layman).

As was once the case in South India (and to some extent still is), there was formerly a close relationship between the Khaṇḍelvāl Jains, as a caste community, and *bhaṭṭāraka*s holding regional seats. These seats have long since become defunct, but when they existed they must have been a strong source of caste identity. Nowadays, the most powerful religious focus for Khaṇḍelvāl Jain identity is the Digambara pilgrimage centre known as Mahavirji (already discussed at the end of Chapter 5) in Karauli District in Rajasthan. Its governing board is dominated by Khaṇḍelvāl Jains.

By far the majority of Jains in the city of Jaipur (the capital of Rajasthan) are Digambara Jains, and almost all are Khaṇḍelvāl Jains. They are considered a trading caste, and many of the city's prominent business families are numbered among them. Most famously, Banjilal Tholia, the man who pioneered Jaipur's well-known gemstone industry, was a Khaṇḍelvāl Jain. However, their most important niche in the city's economic life has been in service occupations. Service as state bureaucrats and functionaries seems to have provided their initial beachhead in Jaipur, and they have continued in this role in the post-Independence era. They have taken up similar jobs in businesses and other organisations, and have also been prominent in such professions as law and chartered accountancy.

The Khaṇḍelvāl Jain origin narrative (here taken from Kāslīvāl, 1989, pp. 64–9) runs as follows. There was once a Rājpūt king of Khandela named Khaṇḍelgiri. (The Rājpūts, considered to be Kṣatriyas, are the martial aristocracy of the region.) He was a Jain (in this version of the story), but his ministers and priests were devotees of the Hindu deity Śiva. A time came when a great plague infected the kingdom and the people began to flee the city. Seeking a solution, the desperate king turned to his Brāhmaṇ advisers who informed him that only a human sacrifice would end the epidemic. The king demurred, but the Brāhmaṇs remained convinced that a human sacrifice was the only answer. Now, it seems that at this very time a group of five hundred Digambara monks had come to Khandela and were meditating in a garden just outside the city. Under cover of darkness, the Brāhmaṇs went there, seized some of the monks and fed them into a sacrificial fire. As a result, the plague struck the kingdom with renewed fury.

When Aparājit Muni, a prominent Digambara *ācārya* of the time, learned of these events, he sent a monk named Jinsen to the scene. Upon his arrival, Jinsen called the city's Jains to a place outside the city where they could take refuge from the plague under the protection of the Jain goddess Cakreśvarī. In this way, all the Jains of Khandela were saved. In the meantime, King Khandalgiri himself had contracted the plague. Having tried all other remedies to no avail, the king journeyed to the city's outskirts to meet with Jinsen at the Jains' place of refuge. He stayed with the monk for seven days, purifying his diet and manner of life, and praying to the Tīrthaṅkara. Simultaneously, Jinsen prayed to Cakreśvarī for the king's recovery. The king was saved, and Jinsen announced that, from that day forward, the king would be under the protection of the Tīrthaṅkara and that the king would publicly proclaim himself to be a Jain.

King Khandalgiri then returned to his palace where, in 44 CE, he and thirteen of his Rājpūt feudatory lords became initiated Jain laymen in the presence of Jinsen and his mendicant disciples. These fourteen Rājpūt noblemen became the ancestors of the original fourteen of the eighty-four patriclans of the Khaṇḍelvāl Jain caste. Thus, the caste was born.

The Osvāls

The Osvāl caste, with its mixed membership of Jains and Hindus, presents an ethnographically somewhat more complex example than the Khaṇḍelvāl Jains. Osvāls are found in large numbers in Rajasthan, Gujarat and wherever business opportunities have drawn them. The caste name derives from the town of Osian, located about twenty-eight miles north of Jodhpur, which the Osvāls claim to be their place of origin. Although they share the same caste name and tradition of origin, the Osvāls of Gujarat and north-eastern Rajasthan have significantly different social institutions and do not inhabit a single social universe.

Those Osvāls who are Jains belong almost exclusively to the Śvetāmbara branch. There is, however, a small group of Osvāls who are affiliated with the nominally Digambara reformist sect known as the Kānjī Svāmī Panth. The Śvetāmbara Osvāl Jains of Rajasthan and Jaipur are divided between the image-worshippers and the two reformist sects – the Sthānakavāsīs and Śvetāmbara Terāpanthīs. In Jaipur, the image-worshipping Osvāls are mostly adherents of the Kharatara Gaccha, but adherents of the Tapā Gaccha also have a temple in the heart of the old city.

The Osvāls of Jaipur are a far smaller community than the Khaṇḍelvāl Jains, but they are far wealthier because they, unlike the Khaṇḍelvāl Jains of the city, are almost entirely in business with a strong caste culture of trade. In that city, they (together with the Śrīmāls, to be discussed momentarily) constitute something of a merchant elite. This is because they and the Śrīmāls have historically dominated the lucrative and prestigious gemstone business. Although a Khaṇḍelvāl Jain invented this business as we see it in Jaipur, the Osvāls' strong traditions of trade enabled them to flourish in the industry to an extent that others could not (Babb, 2013).

In addition, the Osvāls of Rajasthan have historically had a strong relationship with the region's Rājpūt aristocracy, many of whose estates and states they served as bankers and high officials. Members of other trading castes did the same, but the Osvāls seem to have particularly distinguished themselves in this role. Some of these families adopted Rājpūt dress, manners and customs; indeed, some left Jainism to become Vaiṣṇava Hindus. According to one scholar, some of these Rājpūtised Osvāls actually took up a non-vegetarian diet in the late Mughal period, but later returned to vegetarian ways (Mehta, 1999, p. 145).

According Osvāl tradition (here taken from Jñānsundarjī, 1929; see also Meister, 1993; on Jñānsundarjī, see Cort, 2008), in the year 457 BCE a Śvetāmbara Jain *ācārya* named Ratnaprabhsūri, who belonged to the Upkeśa Gaccha, came to the town of Osian, then ruled by a king named Upaldev, with five hundred disciples. He did so at the urging of the Jain goddess Cakreśvarī. The monks settled on a nearby hill, where they fasted for a month. When their fast had ended some of them went into Osian on alms rounds. However, the denizens of the town were meat-eating devotees of a ferocious goddess by the name of Cāmuṇḍā. Because of this, there was no food available that was suitable for Jain mendicants, and Ratnaprabhsūri decided to move on. But Cāmuṇḍā was quite unhappy about this, because not to receive a distinguished Jain *ācārya* with proper hospitality, especially one who had come at the urging of Cakreśvarī, would be disgraceful. She, therefore, asked him to stay for the rainy season retreat, saying that doing so would be highly beneficial. So Ratnaprabhsūri stayed on, which he did in the company of thirty-five of his toughest disciples.

Then, one night, a poisonous snake bit the king's son-in-law (who was also the son of the kingdom's chief minister) as he lay sleeping. The boy appeared to be dead, but when the king and his subjects were conveying the corpse to the burning grounds they were accosted by a Jain mendicant (actually Cāmuṇḍā in disguise) who asked them why they were about to burn a living person. The mendicant then vanished, at which point the people recalled that there was a group of Jain mendicants camped nearby. They took the body there, and the boy was restored to life by being sprinkled with water in which Ratnaprabhsūri's feet had been washed.

Ratnaprabhsūri rejected the king's offer of material rewards and instead suggested that the king and his subjects become Jains. When they agreed to do so, he delivered a sermon. It was long and covered many matters, but among its major themes was the evil of sacrifices and of the hypocrites (here referring mainly to Brāhmaṇs) who promoted and conducted them, and whose actual motive for doing so was to eat the meat. He said further that he would teach them a non-violent type of sacrifice that would burn away the *karma*s that had burdened the soul from beginningless time. Then, sprinkling sanctified powder on the king and his minister and subjects, Ratnaprabhsūri initiated them as lay Jains. In this way, the Osvāl caste was born.

But there remained a serious problem. Cāmuṇḍā had been instrumental in the conversions to Jainism, but she was worried about one important point, which was the animal sacrifices that she so enjoyed. At the time of the conversions she had asked Ratnaprabhsūri whether she would have to give them up, and his answer was ambiguous. As the autumn Navrātrī (a twice-yearly festival of nine days' duration dedicated to the goddess) drew near, the people of the town became seriously worried, because this was a festival at which Cāmuṇḍā would expect animal sacrifice, which they, as Jains, would be unable to provide. Ultimately, Ratnaprabhsūri intervened, with the result that Cāmuṇḍā gave up meat and liquor and became a proper Jain goddess. Ratnaprabhsūri renamed her Sacciyā (true) because she had spoken truthfully when she had told him that a rainy season stay in Osian would be beneficial, and under this name she became the protective goddess of the Osvāls.

Śrīmāls

These are found mainly in Gujarat (where they are called Śrīmālīs) and Rajasthan and, of course, wherever elsewhere their business affairs have taken them. As in the case of the Osvāls, the caste possesses more nominal than sociological reality, and the Śrīmāls of north-eastern Rajasthan (which includes Jaipur) are socially quite separate from the Srīmālīs of Gujarat. The Śrīmāls are mostly Śvetāmbara Jains, although there are significant numbers of Hindus among the Gujarati Śrīmālis. They have a strong business tradition, and occupy the apex of Jaipur's gemstone industry with the Osvāls. The lifestyles of these two castes are very similar, and intermarriage between them has become common in Jaipur, which is probably a consequence of the density of their interaction in the gemstone business.

The Śrīmāls of north-eastern Rajasthan are not only overwhelmingly Śvetāmbara Jains, but virtually all of them belong to the image-worshipping branch. They are also steadfast supporters of the Kharatara Gaccha, and in fact the caste's strong relationship with this mendicant lineage appears to an important foundation of its local corporate identity in Jaipur. The Śrīmāls probably came to the city before the Osvāls, and it was they who dominated the pre-modern jewellery business in the city's early years. It was the Śrīmāls who established the Kharatara Gaccha as the city's dominant image-worshipping mendicant lineage.

In one of its variants, Śrīmāl tradition (here drawn from Rāmlālji, 1910) tells of how the Śrīmāl caste originated in the ancient city of Śrīmāl (now known as Bhinmal, in Jalor District in Rajasthan).

According to this tale, at a time when Lord Mahāvīra was still alive and teaching, the city of Śrīmāl was ruled by a king named Śrīmall, who was a follower of the Vedic religion. He had a daughter named Lakṣmī, for whom he wished to find a husband. He told his Brāhmaṇ advisers that a *svayaṃvara* (a marriage in which a woman chooses her husband at a competition) would be out of the question because such was the beauty of Lakṣmī that the attending kings would surely fight among themselves and kill each other. So the Brāhmaṇs advised him to hold a *aśvamedha* sacrifice (royal horse sacrifice) instead. Not only would he gain merit from the rite, but it would also be attended by many Brāhmaṇs and he would have the benefit of their wise counsel in the matter of his daughter's marriage.

Now it so happens that at this very time Lord Mahāvīra was at Mount Shatrunjaya (a major Jain pilgrimage centre in present-day Gujarat). Knowing that hundreds of thousands of creatures would lose their lives in King Śrimall's proposed sacrifice, he sent his chief disciple, Indrabhūti, with five hundred mendicants to avert the carnage. Hundreds of thousands of Brāhmaṇs had assembled for the rite, and a horse had been taken on a tour of various kingdoms and then brought, ready for sacrifice, to the site of the ceremony. Waiting there were large numbers of other animals who would be burnt in the sacrificial fire. But then, at the very last minute before the ceremony was to begin, Indrabhūti Gautama appeared and delivered a compelling sermon. Conversions to Jainism resulted, and the sacrifice was abandoned. Most of the Brāhmaṇs who were present took the vows of lay Jains. On the Kṣatriya side, the king and 125,000 Rājpūts accepted Jainism and they became the Śrīmāl caste.

Mythic themes and the social identity of Jains

The origin narratives just recounted are but three examples of a very large genre. Such stories deal with the creation – in acts of conversion by Jain mendicants – of Jain castes and also their specific constituent patriclans (especially in the case of the large and heterogeneous Osvāl caste). As a class, the narratives usually link particular mendicant lineages to specific lay Jain communities, the underlying idea being that the lineage of the mendicant who was responsible for the conversion to Jainism has a perpetual tutelary relationship with the lay group he created (Granoff, 1989). In a sense, lay communities are visualised as secondary offshoots of mendicant lineages. These ties have long since lapsed, but the narratives live on because most of them found their way into print by one avenue or another.

The narratives show that if caste (i.e., *jāti*) has no standing at the lofty level of key Jain scriptures, Jainism and caste are deeply interconnected at the level of popular religiosity and social identity. They also indicate that the problem of explaining social origins has a special character in the case of the Jains. The establishment of such groups cannot be traced back to a primordial act of creation, for the Jains have no such concept. So instead, the groups in question – Jain castes – are said to have come into existence because of an act of conversion, with the converts then becoming ancestral to the caste (or a portion of the caste) as it now exists. In other words, the groups become what they are because they accepted a body of doctrine and behavioural norms. They are, one might say, 'cultural' as opposed to 'natural' creations, a point already made about the *varṇa*s created by Ṛṣabha. The ideas and values accepted by converts are themselves uncreated; they exist eternally, to be rediscovered time and again, and without end, by self-enlightened Tīrthaṅkaras, who periodically reinject these truths into human history. The historical present is then connected to these creative moments by the lines of disciplic descent among mendicants through which Jain teachings are reproduced, generation after generation.

As portrayed in these narratives, Jain laity and their caste communities are those who revere, support and are taught by these communities of Jain mendicants. The mendicant lineages socially reproduce themselves by means of disciplic descent, but, of course, they are dependent on lay lineages for physical perpetuation and reproduction. Thus, lay lineages are the source of material support, while lay followers supply (if not always from their own ranks) the initiates who make disciplic descent possible. For their part, the lay communities can physically reproduce themselves, but are dependent on mendicants for social reproduction. This is because Jain teachings – adherence to which was and is central to the social identity of any group as 'Jain' – are, at least in theory, preserved and imparted to the laity by Jain mendicants. It was a sustained relationship with mendicants that enabled Jain Osvāls, to take one instance, to remain Jain Osvāls as opposed to generic Osvāls, who might be Hindu or Jain.

There is one additional point to be made about these myth-histories. Not only are Jain castes created by acts of conversion, but they are also won over from Kṣatriya/Rājpūt backgrounds and crucially, when they convert, they give up their former violent ways. Rājpūts are the warriors and rulers of Rajasthan. They are meat-eating, alcohol-drinking fighters (at least in stereotype and self-image), and they are worshippers of goddesses such as Cāmuṇḍā who demand and consume animal sacrifices. The ethnography of this subject is weak for areas other than Rajasthan, but in Rajasthan the claim of Rājpūt/Kṣatriya origin for Jain castes and clans is widespread and not solely a feature of the three narratives given here.

In assessing the historical accuracy of this claim, it must be noted there is little doubt that the conversion of Hindus to Jains in the region now called Rajasthan was the work of charismatic mendicants, but whether the converts were originally Rājpūts is more difficult to credit. In this region, Rājpūt status is one of the best of all possible social addresses, and so the widespread claim of Rājpūt pedigrees is not surprising. At the same time, the claim does, indeed, have a foundation in the religious culture of Jainism, despite the centrality of non-harm to the Jain outlook. This is because Jainism itself is, as has been seen in this book, arguably a martial tradition.

The martial coloration of Jain religious culture is especially evident in the persona of the Tīrthaṅkara/Jina, the central figure and ego-ideal of Jain tradition at every level. As we know, all Tīrthaṅkaras are born into the warrior class. So important is this tenet that Śvetāmbara tradition emphasises the point by telling of how Indra caused Mahāvīra, when he was still in utero, to be transferred from a Brāhmaṇ womb (in which he had been erroneously conceived) to a Kṣatriya womb in order to keep the rule from being broken. And as we also know, a Tīrthaṅkara is a Jina, and a Jina is 'victor' or 'conqueror'. When a Tīrthaṅkara-to-be is born, his mother has a series of auspicious dream-like visions that portend a crucial choice. The child will grow up to be either a universal emperor or a universal teacher. The matter is decided with the third *kalyāṇaka* – renunciation – which sets him on a course to become a conqueror in the spiritual as opposed to the worldly realm.

This idea is reflected in the conversion narratives that have been examined above, for here also warriors-by-birth renounce the ways of war and turn to another form of conquest.

Another theme of great importance in these origin narratives is the sacrifice, which they present as a catalyst of the group's transformation in conversion. In each narrative, the sacrifice is explicitly linked with lethality (to animals, and in one case to Jain monks). As portrayed in the narratives, the sacrifice is a practice that epitomises all that is wrong about Brāhmaṇs and the social order over which they preside. Each narrative ends with a rejection of that order. In the Khaṇḍelvāl Jain case, human sacrifice, promoted by Brāhmaṇ advisers, brings ruin to the kingdom; the situation is saved only when, under the influence of a Jain mendicant, the Rājpūt king and his henchmen become non-violent Jains. In the Śrīmāl narrative, the potential victims are animals. Jain teachings, imparted by a Jain mendicant, set matters right, and Rājpūt converts become the Śrīmāl caste. In the Osvāl narrative, Ratnaprabhsūri delivers a blast against sacrifice and the Brāhmaṇ hypocrites who promote it at the very point at which he initiates the new Jains. Then we learn that the conversion is not complete until the new converts' clan goddess follows their example by no longer insisting on sacrificial offerings and giving up her non-vegetarian ways.

Rivalry with the Brāhmaṇs and rejection of their blood sacrifices are clearly a timeless theme in Jain religious culture, uniting the present-day social identity of Jains with the ancient Sramana/Brāhmaṇa split from which Jainism emerged into historical visibility.

Jainism and the state

Given the centrality of an ethic of non-harm to Jainism, it is hard to see how Jains could ever be fully at peace with the state. The state has been defined in many ways, but one very useful formulation is Max Weber's, who, in *Politics as a Vocation* (Weber, 1965), famously proclaimed that the state is a political entity that exercises a monopoly on the legitimate use of force within territorial boundaries. If this is indeed what the state is, then obviously Jains must have an ambivalent relationship with it, because its very essence is violence, which Jains deplore. But, at the same time, Jains depend on the state for their personal safety and (in the case of Jains in trade) the creation and maintenance of conditions that make possible the conduct of business. Furthermore, because of their prominence in trade and banking, Jains have often been called upon to bankroll the wars and other projects of rulers. And this, in turn, has frequently opened the way to the appointment of Jains to high ministerial posts.

Historically, all this has meant that Jains have often been deeply engaged in the affairs of states, and this has been so from Jainism's earliest days when Jains were quite successful at attracting royal patronage (as were the Buddhists), which was a major factor in the spread of Jainism in some areas of India. As noted in Chapter 2, royal patronage was an especially crucial factor in Jainism's successful penetration of the subcontinent's South. Jain mendicants appear to have enjoyed royal patronage in Odisha from roughly the second century BCE, and royal patronage was particularly important in Karnataka, where a sustained relationship with royal houses provided the umbrella under which South Indian Digambara communities developed some of their most characteristic institutions. From the eighth century CE, Jainism also found royal patronage in Gujarat, of which the most celebrated instance is Hemacandra's relationship with Kumārapāla. And, in Rajasthan, Jains were deeply implicated in the exercise of state power for centuries, playing key roles as bankers, ministers, stewards and even military leaders in Rājpūt-ruled states and estates.

Jains seem to have shied away from confronting the problem of warfare at the same level of intensity with which they have applied the principle of *ahiṃsā* to other domains of human life. Perhaps this is because of the co-optation of Jains in the machinery of states, or from a sense of the importance of the good opinion of rulers for the welfare of Jains. Or perhaps it has arisen from a sense of the Jains' utter dependence on the state and its institutions, or because of the martial symbolism in Jainism that has just been discussed. Or perhaps all these and other reasons, too, are responsible for the somewhat ambivalent Jain attitude towards warfare. In fact, Jains have served as soldiers and even as highly placed officers in the armies of Indian states. This has been truer of Digambaras than Śvetāmbaras, but it is a fact about both branches. So in light of the Jains' abhorrence of violence, what are we to make of this? Here is P. S. Jaini's response (2000a, p. 11):

> The duty of a Jaina mendicant in this case [self-defense] was quite clear: he must not retaliate in any way and must be willing to lay down his life in order to keep his vow of total nonviolence. For a Jaina layman, however, appropriate conduct was not nearly so clear-cut. There were always situations in which violence would be a last resort in guarding the interests of himself and his community. Unfortunately for the Jaina layman, little comfort was to be found in the Jaina lawbooks on this question, which generally avoided the problem entirely. The Jainas did not presume to legislate on violence that might be perpetrated by a member of society at large. After all, as members of a small minority community, Jainas would have only rarely been called upon to respond to such questions about social violence and would have deferred to the dictates of the worldly standards (*locācāra*) current in the surrounding community.

And moreover, as P. S. Jaini points out in the same discussion (2002a, p. 12), although there exists a Jain doctrine of *virodhihiṃsā*, that is, a doctrine of affirming the legitimacy of meeting violence with violence as a last resort, it is rather half-hearted, and Jain authors have a least refrained from glorifying the violence of warfare.

Perhaps the Jain relationship with the state may be best seen as a special case of a general conundrum that simply cannot be resolved, that of how, without some form of compromise, it is possible to spread the umbrella of non-harming to all living things. Given the world as it is, the rejection of violence cannot be successfully pursued as an absolute; hence the necessity for the social and (as has now been seen) political organisation of virtue in which toleration (if not justification) is extended to at least some of the world's cruel work. And that includes the work of kings.

Afterword

The vast changes that have occurred in post-Independence India, and especially those that took place after the economic liberalisation of the early 1990s, have affected Jains deeply, but for the most part in ways that have less to do with Jainism than with the social and economic classes to which most Jains belong. Jains are generally very well off indeed by Indian standards and have, therefore, been well positioned to benefit from recent economic and political trends favourable to the growing Indian middle and upper-middle classes. Of course, it should be remembered that the Jains are not all well off, despite the stereotype. The Digambara communities of South India are far less affluent than their Śvetāmbara and Digambara co-religionists of the North, and for that matter it is far from true that all northern Jains are wealthy. But again, speaking generally, recent decades have been kind to the Jains, and, to the extent that they must confront the challenges of rapid social and cultural change, they do so in the company of the rest of India, not as Jains in particular.

However, if the focus is shifted from Jains to Jain identity, there has been a recent development of potentially great significance. This has to do with the complex question of the relationship between Jainism and Hinduism. No reader of this book could fail to see that, from the standpoint of its sacred writings and most deeply held truths, Jainism is an internally coherent system of beliefs and practices – sectarian divisions notwithstanding. In this sense, it is indeed clear what Jainism is. Less clear is what Jainism is not, which is the heart of the vexed question of whether or to what extent Jainism is an extension or part of Hinduism. The principal reason for the difficulty has to do with the nature of Hinduism itself, and this, in itself, presents complex issues.

The problem lies in the fact that despite the 'ism' that has come to be affixed to the term 'Hindu', Hinduism is not a single entity but rather a bundle of look-alike religious traditions that are both similar and different from each other in a variety of ways. In fact, the very idea of Hinduism, and thus of Hindu 'identity', is a modern concept. It originated in the nineteenth century, and was a joint product of European scholarship, Indian nationalism and the needs of imperial administrators for a religious category for those among their subjects who were neither Muslims nor Christians.

The question, then, becomes whether or not Jainism belongs to that bundle of traditions. As one might expect, the answer depends on one's perspective. For example, it is sometimes said that acceptance of the authority of the Vedas is a defining characteristic of Hinduism. But if that is so, then there is no way Jainism can be said to be a 'form' of Hinduism for the simple reason that Vedic orthodoxy has been the bête noire of Jainism from its earliest days. However, the 'Vedic' definition of Hinduism is itself highly problematic, because veneration of the Vedas is only one strand – although admittedly an important one – of the mix that is considered to be Hinduism. Then, to add to the complexity, some of these 'Hindu' traditions bear a resemblance to Jainism on one or more dimensions. For example (and as mentioned earlier in this book), the soteriology of the South Indian 'Hindu' tradition known as Śaiva Siddhānta is startlingly similar to Jainism (Davis, 1998, pp. 220–2).

An additional complication in sorting out Jainism's relationship to Hinduism is the fact that the boundary line between Jain and Hindu ritual cultures is in some ways ill-defined. Some events in the Jain ritual calendar are unambiguously Jain in the sense that they are founded on features of Jain belief and are celebrated only by Jains and in a manner that clearly belongs to Jainism's ritual culture. But there are other areas of obvious overlap. An example is the well-nigh universal celebration of Dīvālī by Jains (discussed earlier in this book); while there is a distinctive Jain perspective on the occasion, for the most part Jains celebrate it more or less exactly as their Hindu neighbours do. In addition, Jain temple liturgy has clearly drawn much from Hindu ritual culture. And more, Jains often worship Hindu deities such as Hanumān in the hope that the deity will assist them in some important but worldly matter. In fact, the Jain deity Ghaṇṭākarṇ Mahāvīr is clearly a Jain version of Hanumān.

Without belabouring the point, when all this is put together it seems obvious that whether one stresses the similarities or differences between Jainism and Hinduism finally comes down to what is meant by 'Hindu'. The term itself comes from Persian, and before it acquired its religious meaning in the nineteenth century it was an ethnographic not a religious term. The Hindus were the peoples living beyond the Indus River, and if the term's original meaning is to be emphasised, then any religious tradition indigenous to the subcontinent is 'Hindu'. On the other hand, if the distinctive features of Jainism's belief system are to be highlighted, as in this book, then a good case can be made for its separateness. At any event, much apparent confusion around this issue arises from the frustration of those who insist on finding rigid boundaries where such boundaries do not exist.

To this mix, we must add one point more. The question of religious identity is also a political issue of some importance in India. Hindu nationalism is an appealing ideology to many of India's citizens, and in May 2014 the Hindu nationalist Bharatiya Janata Party (BJP) scored an unprecedented electoral victory and, under the leadership of Narendra Modi, assumed the reins of power in New Delhi. Significant numbers of Jains identify deeply with the political goals and aspirations of Hindu nationalists, and the Hindu nationalists have always favoured the most inclusive concept of Hinduism – that is, one that places Jainism under the umbrella of Hinduism. One may argue that this a 'cultural' rather than 'religious' position, but this distinction does not seem to come naturally to most Hindu nationalists.

Meanwhile, in the previous January had come an important change in the status of the Jains vis-à-vis the Indian republic. This was the granting of legal 'minority religious status' to the Jains by the Congress-led central government then in power. With this, the Jains joined the Muslims, Sikhs, Christians, Buddhists and Parsis as official religious minorities. It should be noted that some thirteen of India's states had already granted minority status to the Jains. The issue of Jain minority status is an old one with a tangled history. Suffice it to say that the National Minorities Commission recommended that the Jains be granted this status in 1993, and that the Union Cabinet finally approved the measure in January 2014.

There has been a great deal of misunderstanding among Jains as well as non-Jains about the effects of this change of status. It does not, as some believe, grant to Jains the reserved government jobs, reserved places in elected assemblies, and reserved seats in government-funded educational institutions granted to disadvantaged groups in Indian society; the Jains are – as a category – obviously not in need of such benefits. The advantages of minority status are in a totally different domain. A major benefit is immunity of Jain temples and trusts from government interference and takeover. Hindu temples have been deprived of such immunity since 1997 when the Supreme Court ruled in favour of a government takeover of the important Vaishno Devi temple in the Indian state of Jammu and Kashmir. An additional benefit is the ability of Jain-managed educational institutions to allow the reservation of up to 50% of seats for Jains and to teach their religion in such schools. Less tangibly, the new status will probably enhance public awareness of the distinctiveness of Jainism and appreciation of the magnitude of its contributions to Indian culture.

To the best of my understanding, this development has been greeted with very mixed feelings by Jains, but it is impossible to attach numbers to this assertion. All that can be said is that Jain responses have differed on the basis of regional, sectarian and class lines. Many are quite pleased by the change and glad to have the aforementioned benefits. Some Jains have opposed the new status out of the mistaken belief that minority status puts Jains in the same category of those disadvantaged (and generally socially stigmatised) groups entitled to government reservations. The new status has nothing to do with such reservations. Some Jains are sympathetic with the Hindu nationalist position, which is that anything that 'divides' Hinduism (as they understand Hinduism) weakens the Hindu nation. Thus, it is hardly a surprise that the Rashtriya Swayamsevak Sangh (RSS) – a militant Hindu nationalist organisation – came out in opposition to minority status for Jains in its annual meeting in March 2014.

In truth, the implications of the change have yet to play out. At this point, it is not possible to know whether, or to what extent or in what ways, the new status will actually affect the manner in which Jains engage with Indian society and politics. Nor is it yet clear that the various benefits of the change will actually prove beneficial in the long run; as is true in most matters of this sort, unintended consequences cannot be discounted. But at the very least, the change inscribes into law the fact of Jainism's independent reality as one of India's greatest religious and intellectual traditions.

References

Agrawal, B. C. (1972) 'Diksha ceremony in Jainism', *The Eastern Anthropologist*, Vol. 31, No. 1, pp. 12–20

Babb, L. A. (1996) *Absent Lord: Ascetics and Kings in a Jain Ritual Culture*, Berkeley, CA: University of California Press

Babb, L. A. (2004) *Alchemies of Violence: Myths of Identity and the Life of Trade in Western India*, New Delhi: Sage Publications

Babb, L. A. (2011) 'Jainism's ethic of non-harm', in Vyas, M. A. (ed.) (2011) *Issues and Ethics in Animal Rights*, Delhi: Regency Publications, pp. 198–211

Babb, L. A. (2013) *Emerald City: The Birth and Evolution of an Indian Gemstone Industry*, Albany, NY: SUNY Press

Balbir, N. (1994) 'Women in Jainism', in Shah, A. (ed.) (1994) *Religion and Women*. Albany, NY: SUNY Press, pp. 121–38

Balbir, N. (2003) 'The A(ñ)calagaccha viewed from inside and from outside', in Qvarnström, O. (ed.) (2003), pp. 47–77

Banks, M. (1992) *Organizing Jainism in India and England*, Oxford: Oxford University Press

Basham, A. L. (1951) *History and Doctrine of the Ājīvikas*, London: Luzac

Bothara, S. (2004) *Ahimsā: The Science of Peace*, Jaipur: Prakrit Bharati Academy

Bothara, S. (2012) 'Gunasthan', in Bothara, S. and Babb, L. A. (eds) (2012) *Ocean of Compassion: Pt. Hargovind Das Sheth's Shri Mahavira Prarthana Shatak*, Jaipur: Prakrit Bharati Academy, pp. 97–102

Bourdieu, P. (1977) *Outline of a Theory of Practice*, Cambridge: Cambridge University Press

Bronkhorst, J. (2000) 'The riddle of the Jainas and Ājīvikas in early

Buddhist literature', *Journal of Indian Philosophy*, Vol. 28, pp. 511–29

Bronkhorst, J. (2013) 'Anekāntavāda, the central philosophy of Ājīvikism?', *International Journal of Jaina Studies* (online), Vol. 9, No. 1, pp. 1–11. Available from URL: https://www.soas.ac.uk/research/publications/journals/ijjs/archive/2013.html (accessed 17 December 2014)

Caillat, C. and Kumar, R. (1981) *The Jain Cosmology*, New York, NY: Navin Kumar Inc.

Carrithers, M. (1989) 'Naked ascetics in southern Digambar Jainism', *Man* (n.s.), Vol. 24, pp. 219–35

Carrithers, M. (1991) 'The foundations of community among southern Digambar Jains: an essay on rhetoric and experience', in Carrithers, M. and Humphrey, C. (eds) (1991), pp. 261–86

Carrithers, M. and Humphrey, C. (eds) (1991) *The Assembly of Listeners: Jains in Society*, Cambridge: Cambridge University Press

Chapple, C. K. (ed.) (2002) *Jainism and Ecology: Nonviolence in the Web of Life*, Cambridge: Centre for the Study of World Religions

Chapple, C. K. (2006) 'Inherent value without nostalgia: animals and the Jaina tradition', in Waldau, P. and Patton, K. (eds) (2006) *A Communion of Subjects: Animals in Religion, Science, and Ethics*, New York, NY: Columbia University Press

Chatterjee, A. K. (1978) *A Comprehensive History of Jainism*, Vol. 1, Calcutta: Firma KLM

Chatterjee, A. K. (1984) *A Comprehensive History of Jainism*, Vol. 2, Calcutta: Firma KLM

Cort, J. E. (1991) 'The Śvetāmbar Mūrtipūjak Jain mendicant', *Man* (n.s.), Vol. 26, No. 4, pp. 651–71

Cort, J. E. (ed.) (1998) *Open Boundaries: Jain Communities and Cultures in Indian History*, Albany, NY: SUNY Press

Cort, J. E. (2001) *Jains in the World: Religious Values and Ideology in India*, New York, NY: Oxford University Press

Cort, J. E. (2002a) 'Bhakti in the early Jain tradition: understanding devotional religion in south Asia', *History of Religions*, Vol. 42, pp. 59–86

Cort, J. E. (2002b) 'A tale of two cities', in Babb, L. A., Joshi, V. and Meister, M. W. (eds) (2002) *Multiple Histories: Culture and Society in the Study of Rajasthan*, Jaipur, Rawat, pp. 39–83

Cort, J. E. (2003) 'Doing for others: merit transfer and karma in Jainism', in Qvarnström, O. (ed.) (2003), pp. 129–49

Cort, J. E. (2005) 'Devotional culture in Jainism: Mānatuṅga and his *Bhaktāmar Stotra*', in Blumenthal, J. (ed.) (2005), *Incompatable Visions: South Asian Religions in History (Essays in Honor of David M. Knipe)*, Madison, WI: Center for South Asia, pp. 93–115

Cort, J. E. (2006a) 'A fifteenth-century Digambar Jain mystic and his followers: Tāraṇ Tāraṇ Svāmī and the Tāraṇ Svāmī Panth', in Flügel, P. (ed.) (2006) *Studies in Jaina History and Culture: Disputes and Dialogues*, London: Routledge, pp. 263–311

Cort, J. E (2006b) 'Installing absence? The consecration of a Jina image', in Maniura, R. and Shepherd, R. (eds) (2006) *Presence: The Inherence of the Prototype within Images and Other Objects*, Aldershot: Ashgate, pp. 71–83

Cort, J. E. (2008) 'Constructing a Jain mendicant lineage: Jñānsundar and the Upkeś Gacch', in Babb, L. A., Cort, J. E. and Meister, M. W. (eds) (2008) *Desert Temples: Sacred Centers of Rajasthan in Historical, Art-Historical, and Social Contexts*, Jaipur: Rawat, pp. 135–69

Cort, J. E. (2010a) 'External eyes on Jain temple icons', in *Initiative for the Study of Material and Visual Cultures of Religion* (online), Yale University. Available from http://mavcor.yale.edu/conversations/object-narratives/external-eyes-jain-temple-icons (accessed 16 January 2015)

Cort, J. E. (2010b) *Framing the Jina: Narratives of Icons and Idols in Jain History,* Oxford: Oxford University Press

Davis, R. H. (1998) 'The story of the disappearing Jains: retelling the Śaiva-Jain encounter in medieval South India', in Cort, J. E. (ed.) (1998), pp. 213–24

Dixit, K. K. (1978) *Early Jainism*, Ahmedabad: L. D. Institute of Indology

Dumont, L. (1970) *Homo Hierarchicus: The Caste System and its Implications*, Chicago, IL: University of Chicago Press

Dundas, P. (1985) 'Food and freedom: the Jain sectarian debate about the nature of the kevalin', *Religion*, Vol. 15, pp. 161–99

Dundas, P. (1991) 'The Digambara Jain warrior', in Carrithers, M. and Humphrey, C. (eds) (1991), pp. 169–86

Dundas, P. (2002) *The Jains* (2nd edn), London and New York: Routledge

Dundas, P. (2003) 'Conversion to Jainism: historical perspectives', in Robinson, R. and Clarke, S. (eds) (2003), *Religious Conversion in India: Modes, Motivations and Meanings*, New Delhi: Oxford

University Press, pp. 125–48

Dundas, P (2007) *History, Scripture and Controversy in a Medieval Jain Sect*, London: Routledge

Ellis, C. M. C. (1991) 'The Jain merchant castes of Rajasthan: some aspects of the management of social identity in a market town', in Carrithers, M. and Humphrey, C. (eds) (1991), pp. 75–107

Flood, G. (1996) *An Introduction to Hinduism*, Cambridge: Cambridge University Press

Flügel, P. (1995–6) 'The ritual circle of the Terāpanth Śvetāmbara Jains', *Bulletin d'Etudes Indiennes*, Vols 13–14, pp. 117–78

Flügel, P. (2006) 'Demographic trends in Jaina monasticism', in Flügel, P. (ed.) (2006) *Studies in Jaina History and Culture: Disputes and Dialogues*, London: Routledge, pp. 312–98

Folkert, K. (1989) 'Jain religious life at ancient Mathura: the heritage of late Victorian interpretation', in Folkert, K. (1993) *Scripture and Community: Collected Essays on the Jains* [edited by J. E. Cort], Atlanta, GA: Scholars Press, pp. 95–11

Glasenapp, H. von (English trans. Gifford, G. B.) (1942 [orig. 1914]) *The Doctrine of Karman in Jain Philsophy*, Bombay: Bai Vijibai Jivanlal Panalal Charity Fund

Glasenapp, H. von (English trans. Shrotri, J. B.) (1999 [orig. 1925]) *Jainism: An Indian Religion of Salvation*, Delhi: Motilal Banarsidass

Granoff, P. (1989) 'Religious biography and clan history among the Śvetāmbara Jains in North India', *East and West,* Vol. 39, Nos 1–4, pp. 195–215

Granoff, P. (1990) (ed.) *The Clever Adulteress and Other Stories: A Treasury of Jain Literature*, Oakville, Ont: Mosaic Press

Granoff, P. (1998) 'Divine delicacies: monks, images, and miracles in the contest between Jainism and Buddhism', in Davis, R. H. (ed.), *Images, Miracles and Authority in Asian Religious Traditions*, Boulder, CO: Westview Press, pp. 121–41

Heim, M. (2004) *Theories of the Gift in South Asia: Hindu, Buddhist, and Jain Reflections on Dāna*, New York and London: Routledge

Hemprabhāśrī (1977) *Śrī Jain Dharm Praveśikā*, Calcutta: Hīrālāl Lūṇyā

Hoernle, A. F. R. (1890) 'The paṭṭāvali or list of the Upkeśa-Gacchha', *Indian Antiquary*, Vol. 19, pp. 233–42

Humphrey, C. and Laidlaw, J. (1994) *The Archetypal Actions of Ritual: A Theory of Ritual Illustrated by the Jain Rite of Worship,* Oxford: Oxford University Press

Jacobi, H. (trans.) (1884) *Gaina Sutras*, Oxford: Clarendon Press

Jain, J. (1975) *Religion and Culture of the Jains*, Delhi: Bharatiya Jnanpith

Jain, K. C. (2010) *History of Jainism* (3 vols), New Delhi: D. K. Printworld

Jaini, J. L. (ed. and trans.) (1974a) *Samayasara (The Soul-Essence) by Kunda Kunda Acharya* (reprint of 1930 original), New York, NY: AMS Press.

Jaini, J. L. (ed., trans. and commentary) (1974b) *Tattvarthadhigama Sutra by Umasvami Acharya* (reprint of 1920 edn, issued as Vol. 2 of the *Sacred Books of the Jainas*), New York, NY: AMS Press

Jaini, P. S. (1977) '*Bhavyatva* and *abhavyatva*: a Jaina doctrine of "predestination" ', in Jaini, P. S. (ed.) (2000b), pp. 95–109

Jaini, P. S. (1979) *The Jaina Path of Purification*, Berkeley, CA: University of California Press

Jaini, P. S. (1980) '*Karma* and the problem of rebirth in Jainism', in Jaini, P. S. (ed.) (2000b), pp. 121–45

Jaini, P. S. (1987) 'Indian perspectives on the spirituality of animals', in Jaini, P. S. (ed.) (2000b), pp. 253–66

Jaini, P. S. (1991) *Gender and Salvation: Jaina Debates on the Spiritual Liberation of Women*, Berkeley, CA: University of California Press

Jaini, P. S. (2000a) '*Ahiṃsā*: a Jaina way of spiritual discipline', in O'Connell, J. (ed.) (2000) *Jain Doctrine and Practice: Academic Perspectives*, Toronto: University of Toronto Centre for South Asian Studies, pp. 1–17

Jaini, P. S. (ed.) (2000b) *Collected Papers on Jaina Studies*, Delhi: Motilal Banarsidass

Jaini, P. S. (2003) 'From nigoda to mokṣa: the story of Marudevī', in Qvarnström, O. (ed.) (2003), pp. 1–27

Javeri, J. S. and Kumar, Muni M. (2008) *Jain Biology*, Ladnun: Jain Vishva Bharati University

Jñānsundarjī, Muni (1929 [Vikram Samvat 1986]) *Śrī Jain Jāti Mahoday*, Phalodi: Śrī Ratnaprabhākar Jñān Puṣpmālā

Kāslīvāl, K. (1989) *Khaṇḍelvāl Jain Samāj kā Vṛhad Itihās*, Jaipur: Jain Itihās Prakāśan Sansthān

Keay, J. (2000) *India: A History*, New York, NY: Atlantic Monthly Press

Kelting, M. W. (2001) *Singing to the Jinas: Jain Laywomen, Maṇḍaḷ Singing, and the Negotiations of Jain Devotion*, Oxford: Oxford University Press

Kelting, M. W. (2009) 'Tournaments of honor: Jain auctions, gender, and reputation', *History of Religions*, Vol. 48, No. 4, pp. 284–308

Laidlaw J. (1985) 'Profit, salvation, and profitable saints', *Cambridge Anthropology*, Vol. 9, No. 3, pp. 50–70

Laidlaw, J. (1995) *Riches and Renunciation: Religion and Economy among the Jains*, Oxford: Oxford University Press

Long, J. D. (2009) *Jainism: An Introduction*, London: Tauris

Marriott, M. (1976) 'Hindu transactions: diversity without dualism', in Kapferer, B. (ed.), *Transaction and Meaning: Directions in the Anthropology of Exchange and Symbolic Behavior*, Philadelphia, PA: Ishi, pp. 109–42

Matilal, B. K. (1981) *The Central Philosophy of Jainism (Anekānta-Vāda)*, Ahmedabad: L. D. Institute of Indology

McGregor, R. S. (ed.) (1992) *Hindi-English Dictionary*, Delhi: Oxford University Press

Mehta, L. (1999) *Caste, Clan and Ethnicity: A Study of Mehtas in Rajasthan*, Jaipur: Rawat

Mehta, S. (2004) *Maximum City: Bombay Lost and Found*, New York, NY: Knopf

Meister, M. W. (1993) 'Sweetmeats or corpses? Community, conversion, and sacred places', in Babb, L. A., Cort, J. E. and Meister, M. W. (eds) (2008), *Desert Temples: Sacred Centers of Rajasthan in Historical, Art-Historical, and Social Contexts*, Jaipur: Rawat, pp. 23–41

Muktiprabhvijay (n.d.) *Śrāvak ko Kyā Karnā Cāhiye?* Vaḍhvāṇ Śahar: Kalyāṇ Sāhitya Prakāśan

Norman, K. R. (1991) 'The role of the layman according to the Jain canon', in Carrithers, M. and Humphrey, C. (eds) (1991), pp. 31–9

Olivelle, P. (1993) *The Āśrama System: The History and Hermeneutics of a Religious Institution*, New York and Oxford: Oxford University Press

Parry, J. (1994) *Death in Banaras*, Cambridge: Cambridge University Press

Peterson, I. V. (1998) 'Śramaṇas against the Tamil way: Jains as others in Tamil Śaiva literature', in Cort, J. E. (ed.) (1998), pp. 163–85

Qvarnström, O. (ed.) (2003) *Jainism and Early Buddhism: Essays in Honor of Padmanabh. S. Jaini*, Part I, Freemont, CA: Asian Humanities Press

Rāmlāljī, Yati (1910) *Mahāvaṃś Muktāvalī*, Bombay: Nirṇaysāgar Press

Reynell, J. (1987) 'Prestige, honour and the family: laywomen's religiosity amongst the Svetambar Murtipujak Jains of Jaipur', *Bulletin d'Etudes Indiennes*, Vol. 5, pp. 313–59

Reynell, J. (1991) 'Women and the reproduction of the Jain community', in Carrithers, M. and Humphrey, C. (eds) (1991), pp. 41–65

Sahgal, S. (1994) 'Spread of Jinism in North India between circa 200 B.C. and circa A.D. 300', in Bhattacharyya, N. N. (ed.) (1994), *Jainism and Prakrit in Ancient and Medieval India: Essays for Prof. Jagdish Chandra Jain*, New Delhi: Manohar, pp. 205–32

Sangave, V. A. (1980) *Jaina Community: A Social Survey*, Bombay: Popular Prakashan

Shah, N. (1998) *Jainism: The World of Conquerors* (2 vols), Brighten and Portland: Sussex Academic Press

Shanta, N. (1997) *The Unknown Pilgrims: The Voice of the Sādhvīs: The History, Spirituality and Life of the Jaina Women Ascetics*, Delhi: Sri Satguru Publications

Sikdar, J. C. (1974) *Jaina Biology*, Ahmedabad: L. D. Institute of Indology

Singhi, N. K. (1991) 'A study of the Jains in a Rajasthan town', in Carrithers, M. and Humphrey, C. (eds) (1991), pp. 139–61

Stark, R. and Bainbridge, W. (1985) *The Future of Religion*, Berkeley, CA: University of California Press

Tatia, N. (trans., ed. and commentary) (1994) *Tattvārtha Sūtra: That Which Is* (by Umāsvāti/Umāsvāmī), New York, NY: HarperCollins

Trautmann, T. R. (1981) *Dravidian Kinship*, Cambridge: Cambridge University Press

Trautmann, T. R. (2011) *India: Brief History of a Civilization*, Oxford: Oxford University Press

Vallely, A. (2002) *Guardians of the Transcendent: An Ethnography of a Jain Ascetic Community*, Toronto: University of Toronto Press

Vallely, A. (2004) 'The Jain plate: the semiotics of the diaspora diet', in Jacobsen, K. A. and Kumar, P. P. (eds) (2004) *South Asians in the Diaspora: Histories and Religious Traditions*, Leiden: Brill

Vatuk, S. (1972) *Kinship and Urbanization:White-Collar Migrants in North India*, Berkeley, CA: University of California Press

Vinayasāgar, Mahopadhyāya (ed. and Hindi trans.; English trans. Lath, M.) (1984) *Kalpasūtra*, Jaipur: Prakrit Bharati

Weber, Max (1965) *Politics as a Vocation,* trans. by Gerth, H. H. and Mills, C. W., Philadelphia, PA: Fortress Press

Wiley, K. L. (2004) *Historical Dictionary of Jainism*, Lanham, MD: Scarecrow Press

Wiley, K. L. (2006a) 'Ahiṃsā and compassion in Jainism', in Flügel, P. (ed.) (2006) *Studies in Jaina History and Culture: Disputes and*

Dialogues, London: Routledge, pp. 438–55
Wiley, K. L. (2006b) 'Five-sensed animals in Jainism', in Waldau, P. and Patton, K. (eds) (2006) *A Communion of Subjects: Animals in Religion, Science, and Ethics*. New York, NY: Columbia University Press, pp. 250–5
Williams, R. (1983) *Jaina Yoga: A Survey of the Medieval Śrāvakācāras* (reprint of 1963 original), Delhi: Motilal Banarsidass
Zimmer, H. (1974) *Myths and Symbols in Indian Art and Civilization*, Princeton, NJ: Princeton University Press

Glossary

Ācārāṅgasūtra	first of the twelve Aṅgas
ācārya	religious preceptor; mendicant leader
adho-loka	the lower world
advaita vedānta	monistic school of Indian philosophy
āgama	scripture
aghātiyā	non-harming *karma*s
*agra pūj*ā	worship before [the image]
Agravāl	trading caste of northern India
ahiṃsā	non-harm or non-violence
ailaka	among Digambaras, the second and highest preparatory stage before becoming a *muni*
ajīva	non-sentient, non-soul
Akṣaya Tṛtīya	Immortal Third, a Jain holy day
ananta	infinite
anekāntavāda	the doctrine of many-pointedness
aṅg pūjā	limb worship
Aṅga	limb; one of the twelve principal Śvetāmbara texts
aṇuvrata	lesser vows taken by laity
aparigraha	non-possession
āpo-kāyika	water-embodied life
Ardhamāgadhī	a Prākrit; language of Śvetāmbara scriptures
arihaṃta (or *arhat*)	one who has achieved omniscience
āryikā	a Digambara nun of the highest level; she cannot become a full *muni*
asaṃkhyāta	uncountable
āsrava	karmic influx
aṣṭaprakārī pūjā	the eightfold worship

aśvamedha sacrifice	royal horse sacrifice
atiśaya kṣetra	pilgrimage place where a miraculous event occurred
ātman	self, soul
avasarpiṇī	declining epoch
āvaśyaka	daily essential duties of mendicants
āyambil	a fast in which only sour foods are taken
ayogya-kevalin	motionless *kevalin*
bandha	karmic bondage
Baniyā	member of a trading caste
bhakti	devotion
bhaṭṭāraka	non-peripatetic and clothing-wearing religious functionary (Digambara)
bhāva	spiritually beneficial feelings
bhavyatva	an innate capability to be liberated
bhogabhūmi	land of enjoyment
Brāhmaṇ	the *varṇa* of priests and teachers
caitanya	consciousness
caityavāsī	temple- or monastery-dwelling monks
cakravartin	universal emperor
caturmāsa	four-month rainy-season retreat
caturvidha saṅgha	fourfold order of Jain society
dādābāṛī	garden of the *dādā*s (grandfathers); a shrine where the Dādāgurus are worshipped
Dādāguru	one of four venerated monks of the past belonging to the Kharatara Gaccha
dāna	alms, a charitable gift; a merit-generating gift
darśana	auspicious seeing of a deity or august personage; a school of philosophy
darśana-pratimā	the first *pratimā* – the stage of 'correct viewpoint'
Daśalakṣaṇaparvan	Digambara equivalent of the Svetāmbaras' Paryuṣaṇa
dharma	sacred duty, righteousness; religion
Digambara	sky-clad; the branch of Jainism whose monks are nude
dīkṣā	initiation of a mendicant
Dīvālī	Hindu festival also celebrated by Jains
Dṛṣtivāda	the lost twelfth Aṅga
duṣamā	unhappy

dveṣa	aversion
gaccha	mendicant lineage or clan
gaṇadhara	the Tīrthaṅkaras' chief disciples
gati	destiny class; one of four categories of living things
ghātiyā	harming *karma*s
gocari	alms rounds; a cow's grazing
guṇasthāna	the fourteen stages of spiritual advance towards liberation
*guṇavrata*s	subsidiary vows
gupti	curtailment; restriction
guru	spiritual teacher
Indra	the king(s) of the gods, of which there are sixty-four
Indrabhūti Gautama	Mahāvīra's chief disciple
Indrāṇī	the consort of an Indra
Jambū Dvīpa	Black Plum Continent; the central island of the terrestrial disc
jāti	caste
Jina	victor; one who has defeated all desires and aversions; a synonym for Tīrthaṅkara
jīva	life, soul
Jyotiṣas	solar, plenatary and stellar deities
Kalpasūtra	a text in the Śvetāmbara canon
kalyāṇaka	one of the five auspicious events that occur in the final lifetime of a Tīrthaṅkara: conception, birth, renunciation, attainment of omniscience and final liberation
karma	action that affects the subsequent destiny of the actor; in Jain doctrine, a material substance that adheres to the soul
karmabhūmī	land of action; land of endeavour; those areas where suffering occurs, where humans must work for subsistence, and where liberation is possible
karmayuga	age of action; age of endeavour; the temperal equivalent of *karmabhūmī*
karuṇā	compassion
kāyotsarga	a bodily position expressing abandonment of the body

kevalajñāna	omniscience
kevalin	omniscient being
Khaṇḍelvāl	name taken by North Indian castes tracing their origin to the ancient city of Khandela
Kharatara Gaccha	a mendicant lineage among Śvetāmbara Mandirmārgī Jains
Kṣatriya	the *varṇa* of warriors and rulers
kṣetrapāl	guardian deities of temples
madhya-loka	the middle world
mahāpūjā	a major congregational rite of worship
Mahāvideha	a terrestrial zone, half of which is always *karmabhūmi*
mahāvratas	the five great vows of a Jain mendicant
Mandirmārgī	Śvetāmbar Jains who worship images in temples; same as Mūrtipūjaka
mantra	sacred verbal formula
mithyādṛṣṭi	the first *guṇasthāna*; the stage of false views
mokṣa	liberation
mokṣa mārga	path of liberation
muhpattī	mouth cloth
muni	mendicant
Mūrtipūjaka	Śvetāmbar Jains who worship images in temples; same as Mandirmārgī
naivedya	food offered in worship
Namaskāra mantra	expression of homage to the five *parameṣṭhin*s
nandīśvara Dvīpa	Isle of Bliss, the eighth concentric island of the terrestrial disc
nigoda	lowest form of life
nirjarā	getting rid of *karma*
Osvāl	North Indian trading caste tracing its origin to Osian in Rajasthan
pāpa	demerit; insalubrious *karma*; opposite of *puṃya*
(five) *parameṣṭhin*(s)	the five supremely worship-worthy beings saluted in the *namaskāra mantra*: *the arihaṃta*s, the *siddha*s, the *ācārya*s, the *upādhyāya*s and the *sādhus*
Pārśva	the twenty-third Tīrthaṅkara of our declining epoch
*parva*s	holy days
Paryuṣaṇa	a period of religious observances taking place

	during the rainy-season retreat; Śvetāmbara
Prakrit (*prākṛt*)	any of several ancient Indo-Aryan vernacular languages
pratikramaṇa	ritual of repentence
pratimā	successive stages of renouncing the world
pratiṣṭha	image consecration ceremony
pratyeka śarīra	plants in which a single soul inhabits a single body
pṛthvī-kāyika	earth-embodied life
pudgala	atom
pūjā	worship
puṇya	merit; salubrious *karma*; opposite of *pāpa*
Pūrva	fourteen canonical texts, now lost
Rājpūt	the martial aristocracy of Rajasthan and other areas of northern and central India
ratnatreya	the three jewels of Jain teachings: right faith, right knowledge, right conduct
Ṛṣabha	the first Tīrthaṅkara of our declining epoch
sādhāraṇa śarīra	plants with multiple souls in a single body
sādhu	male mendicant
sādhvī	female mendicant
Śaiva	worshipper of the Hindu deity Śiva
sallekhanā	ritualised death by self-starvation
samavasaraṇa	the universal assembly of deities, humans and animals who have come to hear a Tīrthaṅkara's preaching
sāmāyika	attaining equanimity; a ritual of meditation for laity
samiti	circumspection
saṃsāra	the cycle of rebirth
saṃvara	stopping karmic influx
saṃvegī sādhu	fully initiated mendicant
samyak cāritra	right conduct
samyak darśana	correct view; faith in the Tīrthaṅkaras' teachings
samyak jñāna	right knowledge
saṃyama	restraint
santhāra	ritualised death by self-starvation
sayoga-kevalin	a still-embodied omniscient being
siddha	a liberated soul
siddha loka / śilā	abode of liberated souls

śrāvaka	'listener'; Jain layman
śrāvakācāra	genre of mendicant-authored books on lay discipline
*śrāvaka-pratimā*s	stages of world renunciation for laity
Śrīmāl (or Śrīmālī)	trading caste of northern and western India
Sthānakavāsī	reformist Jain branch who do not worship images in temple
Śūdra	the lowest *varṇa* class whose traditional duty was 'serving' the upper three classes
sukha	bliss
suṣamā	happy
Śvetāmbara	'white clad'; the branch of Jainism whose mendicants wear white clothing
Tapā Gaccha	mendicant lineage among Śvetāmbara Mandirmārgī Jains
tapas	austerity; ascetic practice
tejo-kāyika	fire-embodied life
Terāpanth	a name held in common by two reformist branches, one Digambara and the other Śvetāmbara
tīrtha	ford, place of pilgrimage
Tīrthaṅkara	one who establishes a 'ford' (as in a ford across a river); an omniscient teacher; synonymous with Jina
tīryañca	the destiny class of animals and plants
trasa	mobile and two-sensed forms of life
trasa nāḍī	chimney-like vertical shaft in the cosmos to which mobile forms of life are restricted
Upāṅga	a group of texts subsidiary to the Aṅgas
upāsaka	formal term for a Jain layperson
upavāsa	fast
ūrdhva-loka	the upper world
utsarpiṇī	ascending epoch
Vaimānika	deities who occupy the levels of heaven
Vaiṣṇava	worshipper of the Hindu deity Viṣṇu or one of his forms
Vaiśya	the *varṇa* class associated with trade and banking
vandana	ritual of veneration
varṇa system	ancient idealisation of society as consisting of

	four classes: Brāhmaṇ, Kṣatriya, Vaiśya and Śūdra
vāyu-kāyika	air-embodied life
*vimāna*s	aerial vehicles used by the Vaimānika deities
vīrya	energy, strength
vītarāga	devoid of desire and aversion
vrata	vow
yati	Śvetāmbara mendicant whose vows are less onerous than those of full mendicants
yātra	pilgrimage
yojana	a unit of distance; about eight miles
yoni	womb, female genitalia; birth situation; a category of living thing roughly equivalent to species

Further Reading

In what follows I have indicated a few starting points for further reading. Many are works already cited and (sometimes suggested) in the text of this book. There are many excellent choices for background reading in Indian history. Among the best are Keay (2000) and, much briefer but highly readable, Trautmann (2011). P. S. Jaini (1979) and Dundas (2002) are the leading general works on Jainism, but there are many other good general studies, highly varied in content and approach, including Glasenapp (1999), J. Jain (1975), Long (2009) and Shah (1998). In addition, Wiley's (2004) superbly researched historical dictionary is an invaluable general resource for both novice and advanced students of Jainism. A good starting point for the history of Jainism would be the relevant chapters of Dundas (2002), and background material on the evolution of non-Jain traditions can be found in Flood (1996), to name but one possibility. More specialised sources on Jain history are Chatterjee (1978, 1984), Dixit (1978) and K. C. Jain (2010). The basics of Jain belief have never been more lucidly set forth than in P. S. Jaini's writings (especially 1979, 1980), but also see Matilal (1981) on *anekāntavāda* and Glasenapp (1942) for a highly detailed survey of *karma* theory. Two excellent accounts of the mendicant lifestyle are Shanta (1997) and Vallely (2002). Studies of lay life among the Jains include Babb (1996), Cort (2001), Kelting (2001) and Laidlaw (1995); Williams (1983) provides a survey of the classical manuals of discipline. Jain understandings of the universe and the world of living things are discussed in most general works on Jainism. Those wishing to pursue the matter in greater detail would do well to go directly to Umāsvāti's *Tattvārtha Sūtra*; I suggest Tatia's translation (1994). A beautifully illustrated survey of Jain cosmography is Caillat and Kumar (1981), and although the main focus of Cort (2010b) is iconography, this work contains excellent material on the Jain cosmic vision. Interesting discussions of Jain views of the world of life and the potential for bridges to modern environmentalism can be found in Chapple (2002, 2006). For the social organisation of Jain communities, see Babb (2004), Carrithers (1991), Ellis (1991), Sangave (1980) and Singhi (1991). While not dealing with Jains in particular, Vatuk (1972) sketches out kinship and marriage patterns typical of the social milieu to which many North Indian Jains belong. Readers seeking a good introduction to Jainism's rich and centuries-old story literature will find it in Granoff (1990).

Those interested in pursuing online resources should turn first to the website of the Federation of Jain Associations in North America (JAINA) (www.jaina.org), which supplies a comprehensive list of web resources conveniently organised by category.

Index

Page numbers in *italics* denote illustrations, in **bold** glossary entries